当代建筑师系列

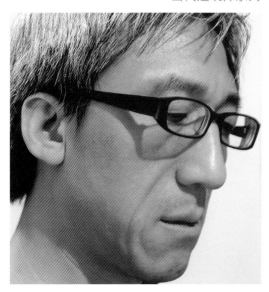

王 昀
WANG YUN

方体空间工作室 编著

中国建筑工业出版社

图书在版编目(CIP)数据

王昀/方体空间工作室编著.—北京:中国建筑工业出版社,2012.6
(当代建筑师系列)
ISBN 978-7-112-14306-1

Ⅰ.①王… Ⅱ.①方… Ⅲ.①建筑设计-作品集-中国-现代②建筑艺术-作品-评论-中国-现代 Ⅳ.①TU206②TU-862

中国版本图书馆CIP数据核字(2012)第091505号

策划编辑:陆新之
责任编辑:刘 丹 徐 冉
责任设计:赵明霞
责任校对:王誉欣 赵 颖

感谢山东金晶科技股份有限公司大力支持

当代建筑师系列
王昀
方体空间工作室 编著
*
中国建筑工业出版社出版、发行(北京西郊百万庄)
各地新华书店、建筑书店经销
北京嘉泰利德公司制版
北京顺诚彩色印刷有限公司
*
开本:965×1270毫米 1/16 印张:11 字数:300千字
2012年11月第一版 2012年11月第一次印刷
定价:98.00元
ISBN 978-7-112-14306-1
(22364)

版权所有 翻印必究
如有印装质量问题,可寄本社退换
(邮政编码100037)

目录

王昀印象	005-006
王昀论文	007-007
空间经验与意识空间的投射	008-012
王昀设计作品	013-013
白色方体空间设计操作过程中的思考	014-017
善美办公楼门厅增建	018-023
60平方米极小城市	024-031
庐师山庄 A+B 住宅	032-053
庐师山庄会所	054-067
百子湾小区中学	068-073
百子湾小区幼儿园	074-083
石景山财政培训中心	084-091

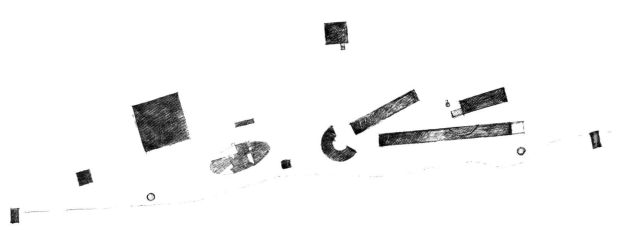

茵莱玻璃钢门窗制品有限公司办公楼	092-101
国轩国际研发中心商业设施	102-105
柔软住宅	106-107
北京万象新天150平方米住宅改造	108-111
苏家坨住宅区小学	112-115
私宅	116-117
鄂尔多斯康巴什第六中学校设计	118-119
无锡天鹅湖会所 A+B 设计	120-123
西溪学社	124-141
鄂尔多斯康巴什住宅组团设计	142-145
王昀——几何抽象与冥想	146-149
对谈——情感、记忆与历史意识	150-154
王昀访谈	155-157
出版物一览	158-159
作品年表	160-171
王昀简介	172-173

CONTENTS

Portrait	005-006
Thesis of Wang Yun	007-007
The projection of space experience and awareness	008-012
Works by Wang Yun	013-013
The Thinking during the Design of White Cube Space	014-017
Annex Foyer of Shanmei Office Building	018-023
"60 Square Meters" Minimum City	024-031
The Lushi Hill A+B House	032-053
The Lushi Hill Club	054-067
BaiziBay District Middle School	068-073
BaiziBay District Kindergarden	074-083
Shijingshan Financial Training Centre	084-091
Inline FRP Doors and Windows Products Co., Ltd. Office Building	092-101
Commercial Facilities of Guoxuan International R&D Centre	102-105

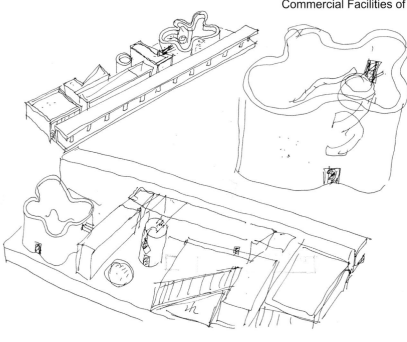

Soft House	106-107
Wanxiangxintian 150 Square Meters Residential Reformation	108-111
Sujiatuo Residential District Primary School	112-115
Private Villa	116-117
Ordos Kangbashi NO.6 Middle School	118-119
Wuxi Swan Club A + B	120-123
XiXi Learning Community	124-141
Ordos Kangbashi Residential District	142-145
Wang Yun——Geometric Abstraction and Meditation	146-149
Discussion——Emotion, Memory and the Sense of History	150-154
Interview	155-157
Publication List	158-159
Chronology of Works	160-171
Profile	172-173

王昀印象

文/黄元炤

王昀，1962年出生，1985年毕业于北京建筑工程学院（现北京建筑大学），后留校任教，20世纪90年代初赴日本东京大学留学，1995年取得硕士学位，1999年获博士学位，他曾用10多年的时间持续地研究世界聚落，21世纪初回国，进入北京大学任教并成立工作室，开始其个人在国内的建筑实践工作，他的工作室取名为方体空间。

聚落研究，已是王昀于中国当代建筑界最明显的表征与品牌形象。他认为民居研究和聚落研究的对象都是一样的，民居在聚落之中，聚落是由民居所构成的整体的、集合状态的呈现。聚落研究，就是对聚在一块儿的民居的研究。民居研究是一个近景的研究，是把视点落在民居上所进行的细部观察，而聚落研究是一个远景的研究，是将视线放远，在整体上对民居的集合状态进行观察与把握。由于他曾有过民居研究和聚落研究的经历，因此，他的聚落研究总在远景与近景之间进行着。而这样的近景和远景之间所形成的摇摆与模糊的作用力，似乎也体现在王昀日后的设计思考与作品成形过程中。王昀关注原生聚落，也关注现代主义建筑思想，一个是传统的面向，一个是现代的面向。这两者在历史、文化、类型、语言方面都存在极大的差异性，而王昀似乎想要摇摆并游走穿越于传统与现代这两者之间，有时共同呈现，有时此多彼少，表面上看去似乎有一种时空上的错乱，有时分离又有时并置，他想把时间轴拉到一个极致的状态，想在现实的空间中创造一个传统与现代的并置关系，而这样的创造还一直持续着。

在王昀看来，这个时空上的错乱状态还与他从小到大的经历及到处探访建筑与聚落有关，他探访不同的环境，接触不同的地方，看不同的东西，而这些东西也杂乱地作用于他了。而王昀的设计作品，则是把所有错乱的集合体经过压缩后拿出的，是很多积累后的一个整合体，看似简单、实则复杂。这种从复杂到简单的转化，是由他自己内在的极致体验与沉淀后的投射与浓缩的结果。而这种投射与浓缩，实际上产生于他那反复于个人意识中的具象与抽象之间的流转。另外，他希望他的建筑是没有性格的，可以悠游于各个位置，又因为之前错乱的复杂作用，所以，他想把建筑简单化、单纯化，回归到一个白色的盒子。

北京庐师山庄会所与别墅，是王昀回归"白盒子"的代表性作品，在这个项目中可以看到现代主义的建筑思想，一种关注纯粹的空间，带有抽象与体验的设计倾向。他所强调的抽象空间，完全是站在20世纪现代主义起始点的思考，就是把人的理性思考回归原点并做到纯粹化的极致。人究竟可以思考到什么样的边缘，可以抽象到什么样的程度是他在作品中的追求和体现。因此他用抽象语言表述出作品内在深层的真实性，在几何体块中创造出内部游走与想象的空间体验过程，而这个空间体验，就是他经过意识的沉淀之后的一个投射结果。而当作品中的空间抽象与体验发挥到极致后，设计者的角色在空间中顿时抽离、消失、隐藏或退位，与观察者同时被并为第三人，成为一个观赏者，而空间本体则跃升为第一人，成为主角，展现着内部富有情结般的路径，并经由多重视点、多重形式的转换而形成幻象，以唤起观赏者的某些记忆，空间完全在表达着建筑最本质的东西。

王昀的作品从北京庐师山庄会所与住宅，到北京百子湾中学，再到杭州西溪湿地创意会所——观察这几件代表性作品，从外部看带有一种一致性，即将建筑简单与单纯化，回归到"白盒子"，并将物体还原至某几个几何形体，由基本的几何语言构成总体的形态。而这些一致性统一后的结果实际上在追求着一种对纯粹空间的探索。同时王昀所设计的每件作品还在这种一致性的基础上维持着差异性、细微的变化、偏离的微差。如北京庐师山庄住宅与北京百子湾中学，王昀将视点均放在合院的抽象上，这两栋建筑内部都框出一个虚拟的天井，但是天井的尺度和空间的性格特征截然不同，而实际上这样的合院抽象化本身还切中和关注到中国性问题的命题。

体验与感受，是王昀设计的作品中特别强调的两点，而如何给人体验与感受是需要事先去设定好场景与意境的，且场景与意境都是夹杂在使用功能前提之下而进行的。场景与意境的设定就是空间的整体设计，而空间也是20世纪现代主义建筑思想最根本的起点，也是王昀作品中所要展现的重点。观察王昀设计的空间，感觉他总想要通过空间程序的编排去刺激人达到一种想象，除了唤起观赏者自己的某些记忆与想象外，他也企图给予观赏者一种设计者本身想要传达的想象。如杭州西溪湿地创意会所，王昀试图通过散落的几何体块，创造出想象的极大自由度，不管是想象过往的聚落，还是想象某个记忆中的场景，抑或是想象江南水乡的白墙黑瓦，都是通过体验丰富与抽象的环绕性空间，在走动的过程中，让视觉达到无限的延伸，使抽象氛围在水平与垂直之间延展，而在浓荫茂密的植被中的建筑，也开始在人的视野中消散与模糊。这样的做法，一方面建筑退隐自然背后，给予场地基本的尊重，一方面建筑逐渐地冲破了既有形式，离散于现实与印象之间。

王昀设计中的空间抽象与体验，刺激人的想象，也与人的动作相关联，而动作的过程本身实际上与设计流线、设定路径有关。观察王昀设计的空间中常会使用坡道，而坡道本身实际上起到拉长行进路径的作用，他力图用它去创造一个过程，开始是缓慢的，想象也是不全的，然后从缓慢到停留，从一个场景转换与衔接到另一个场景，片刻停留去品味空间氛围，同时脑袋中的想象强度又会伴随这种移动而不断地扩大和展开，就会猜想到或许是王昀想要传达的空间意识，体会到来回反复的主客体意识接触和交融。这样从缓慢到停留，再到缓慢，再到停留，来来回回地重复着，人常处于一种半想象与全想象的空间抽象的体验思绪之中。

纵观王昀的设计，空间抽象性从没消失过，体验、感受与想象依然存在着，但他还是努力探索和尝试空间上的变化与探索性的转型。而他所做的这一切最终的追求是试图将空间抽象化发挥到极致，这一点是毋庸置疑的。如果从建筑学的视角来看，他的作为是正统的，从建筑史学的角度来看，王昀始终坚持在纯粹空间抽象性的探索脉络上，丝毫没有偏移过。

Portrait

By Huang Yuanzhao

Wang Yun, born in 1962, graduated from Beijing University of Civil Engineering and Architecture and started teaching there in 1985. In the early 1990s, he went to Japan and gained his master's degree in 1995 and Ph.D. in 1999 from the University of Tokyo. He has spent 10 years in the research of global dwelling settlement. He came back to China at the beginning of the 21st century and teaches at Peking University. His design office is named Atelier Fronti.

Research on Settlement is Wang Yun's distinct personal characterization in Chinese contemporary architectural field. He believes that research object of residence and settlement are the same. Residence is included in settlement, and settlement presents assembling state of residences. Settlement research focuses on observation of residence from an overall point of view, while residence research focuses on architectural details. Because of his research experiences on both residence and settlement, Wang Yun keeps moving between distant view and close shot. This moving also appears in his works and design processes. Wang yun concerns both native settlement and modern architecture, which are different greatly in history, culture, type and language. It seems Wang Yun is walking between tradition and modernity. Sometimes, he wants to present both factors equally. Sometimes one is more than the other. Although seemingly spatial disorder, he wants to pull time axis to an ultimate state and form a juxtaposition of tradition and modernity at the same time. Such creation has always been continued during his work.

In Wang Yun's opinion, the disordered state of space and time related to his own experiences. He has travelled to many places, visited multiple buildings and settlements, and exposed to different environments. All these experiences have significant influences on him. His design works are expressions of all these experiences, and are integrated bodies with a lot of accumulations, seemingly simple but complex actually. This process from complex to simple is the result of precipitation of his inherent experiences. It is produced during his hover between concretization and abstraction. In addition, he hopes his buildings are without characters and capable of suiting each location. With former complex influences, he wants to simplify architectures and return them to white boxes.

Lushi club and villas in Beijing present his returning to "white boxes". He focuses on pure space with abstract design tendencies. Based on the starting point of 20th century modernism, Wang Yun stresses abstract space and pushes original rational thinking to a ultimate point. He tries to pursue and reflect the edge of people's abstract thinking. Therefore, he expresses authenticity in abstract way and creates experiences of walking and imaginations with geometry blocks. These experiences come from his consciousness precipitation. After bringing abstraction and experience to extremeness, designer suddenly disappears out of space and hides into space, and becomes a viewer while space itself becomes the protagonist, showing internal complex paths and imaginations come from the conversion of multiple forms. As a result, memories of the viewers are evoked and space expressions become the most essential part of the architecture.

Wang Yun's representative works include Lushi club and villas in Beijing, Baiziwan primary and secondary schools in Beijing and Xixi Wetland learning community in Hangzhou. All these buildings are simple white boxes whose shapes are formed by basic geometric shapes. They share the character of simplicity. The consistency of the buildings is actually exploration for pure space. All of Wang Yun's buildings have slight differences on the basis of consistency. For example, Wang Yun focuses on the abstraction of courtyard in Lushi Villa and Baiziwan Middle School. Wang Yun focuses on the abstraction of courtyard, both of the two buildings have virtual patio inside. But scales and characters of space are totally different. In addition, this abstraction also related to the issue of the Chinese culture.

Wang Yun's works stress setting scene and create atmosphere on basis of function to bring visitors special experiences and feelings. Setting scene and creating atmosphere means overall design of space. Space Is not only the start of modernism architecture design in 20th century but also Wang Yun's attention on architecture. Wang Yun tries to evoke visitors' imagination and memories as well as express his own imagination through arrangement of spaces. In Xixi Wetland learning community, Wang Yun hope to create the limitless imagination through separated geometry blocks. It reminds you of traditional settlement, a scene in your memory or the atmosphere in South China. During the process of walking, your vision can reach the unlimited extension in the abstract atmosphere. Buildings in the shade of dense vegetation are dissipated and become vague. In this way, the buildings are not only incorporated in environment but also different from existing constructions.

Abstraction and experience of spaces in Wang Yun's design also associate with people's movement decided by organization of paths in the building. Ramp is used by Wang Yun to extend paths and create a process which maybe slow and lack of imagination at the beginning, but with transfer of scenes, people can enjoy the space as well as expand their imaginations. Such a process of slowing down and staying repeatedly provides people an abstract experiences with imaginations.

Space abstraction appears in all of Wang Yun's works; space experiences, feelings and imaginations never disappear. But he also makes efforts to explore space changes and transformations. There's no doubt that his ultimate goal is to maximize space abstraction. From the perspective of architecture, his work is orthodox; from the perspective of architectural history, he adheres to the context of exploring abstraction of pure space.

王昀 论文   Thesis of Wang Yun

庐师山庄A+B住宅

庐师山庄A+B住宅   The Lushi Hill A+B House

空间经验与意识空间的投射　The projection of space experience and awareness

文/王昀
By Wang Yun

一、两块玻璃与两个投射

（一）丢勒版画中的玻璃

在丢勒的版画中，有几幅知性色彩浓厚的版画。它们记录着文艺复兴时期画家绘画时的工作场景，描述着绘画工作的过程。版画中的一幅，描绘的是一位画家正在将空间中的对象物绘制到平面画板中的情景。从图中我们可以看到画家正使用一根绳索，以确认点的方式，将一个三维物体转换到二维平面上。为寻找三维与二维之间的相互关系，以确定三维物体中的点在二维平面中的点的对应位置，绳索的本身成为三维物体与二维平面图像之间保持相互关系的维持物。这种维持物一方面表示着三维物体投射到二维平面过程中的投射轨迹，另一方面也指示着空间与平面之间相互点的对应位置。如果我们将绳索这个表示着三维空间物体向二维平面投射轨迹的维持物去掉，尽管我们看不到轨迹(绳索)本身的存在，但我们仍不难理解物体从三维空间向二维空间投射的结果和事实。一旦我们明了了这一点，并以如此的视点来看待三维空间与二维平面之间的相互关系时，便可以得到下面的结论：

亦即"二维平面中的像是三维空间中的物在二维平面上的投射"。

丢勒本人是一位画家的同时又是一位谙熟数学和几何学的学者，他笔下画家的工作内容，或许不仅仅只是单纯地记录从三维物体投射到二维平面中的点与线的位置，而更或许是在探讨着如何将三维空间中的物体投射和转化到二维平面中的画法几何。

同样引起我极大兴趣的是在这几幅版画中均存在着一个共同的装置，就是在每个画面中都具有的一个如窗子般的框。而这个"框"的本身，实际上就是二维平面的界面和限定范围，也是三维空间中的切片和断面。版画中的"框"实际上又是探索如何将三维的立体世界投射到二维平面世界过程中的富有意味的投射装置和研讨器。我们姑且称之为"空间投射器"。这个投射器的表现形式是透明的玻璃窗。当观者将视点固定在玻璃上，透过玻璃看到其背后的三维物体时，三维物体本身就成为玻璃上的像，并且这个像本身就是二维平面的存在。当将投射在玻璃上的三维的像加以记录，并准确地将其加以描绘时，所截取的像的本身就是三维物体在二维平面上的投射结果，并且二维平面上的像与三维空间中的物之间还同时保持着一种相互对应的关系。

如此透过一个透明的二维界面去看三维世界的方法和视点，就是我们所说的"透视"。很显然所谓"透视"就是"透过去视"，即透过一个透明的二维界面去视，去视一个三维世界投射在二维透明玻璃上的像。而将投射在二维透明界面上的像加以记录以及对另一个三维世界投射的轨迹进行记录的几何学手段就是我们常见的画法几何。

由于对这种投射关系的研究以及相应表现手段的成熟，反过来出现了一

1. Two pieces of glasses and two projections

(A) Glass in Dürer's painting

Some of Dürer's engravings recorded the scene of Renaissance painters painting. One of those paintings depicts that an artist is drawing objects on drawing board with a rope, which is used as a link between three-dimensional objects and two-dimensional surface. The link stands for projection trace of the translation; Also it presents affiliation points in space and on the plane. Even without rope, people can easily understand the results and facts of the projection. Based on that, we can get the following conclusions:

"Objects in two-dimensional plane are the projection of objects in three-dimensional space".

Dürer himself was a painter as well as a scholar of mathematics and geometry. His paintings not only record points and lines projected objects from three-dimensional space to two-dimensional plane, but also explore descriptive geometry to realize this projection.

A device appears in all these engravings arouses my great interests. The window-like frame in the interface of plane playing the role of restricting the range of the two-dimensional space, which is the cross-section of the three-dimensional space. The "frame" is actually the device in exploring the projection from three-dimensional world to two-dimensional world. We can call it "space projector" (with the form of transparent glass window). Observing objects through the projector, three-dimensional objects become two-dimensional existence. When we try to portray these existences accurately, we get projected results of the three-dimensional objects in the space. Mutual corresponding relationship exists between two-dimensional figure and three-dimensional object.

The method above is so called "perspective", which means to observe object through a transparent two-dimensional interface, and to observe projections of a three-dimensional world. The whole process of recording the two-dimensional figure and projection is called descriptive geometry.

Further projection research created a new way to present space and three-dimensional projection. As a result, artists after Renaissance painted "window frame" to express three-dimensional space. In their eyes it means perspective frames. Objects seen through the frame is the projection of three-dimensional object. But regards the results of two-dimensional

个新的空间表现方式，就是利用二维平面去表现三维空间，即用二维平面表现三维空间的投射的结果。也正是因为如此，观察在"窗"上所投射世界的观察角度和表现方法后来一直成为文艺复兴之后在平面中表现三维空间和窗景的写实绘画的基本根基。即：画框=透视的窗框。透过画框观察到的对象也就是现实三维空间的对象在画框中投射的事实。但是这种仅从二维平面的投射结果就去判断三维世界全部的观念和思想，实际上对于空间维度的认识和扩张判断都有着巨大的局限性和危害性。

（二）杜尚的玻璃

上述这种空间的认识和表现在20世纪初由于科学的进步而得到重新审视。与这种空间问题思考相呼应的是画家杜尚所做的一块以"新娘"为命题的大玻璃，这是一块与丢勒的玻璃窗极为相似的大玻璃窗，其制作的目的是试图寻找一种新的透视法，即探索寻找出一个由四维世界向三维空间投射的透视法则。

既然二维平面中的像是三维空间中的物在二维平面上的投射，那么我们这个三维世界的存在，不正是一个四维世界的空间投射的结果的存在吗？按照这样的理解，杜尚将投射到二维玻璃上的三维世界不再视为二维平面的存在，而是三维本身的存在。因此玻璃窗上（杜尚的所谓的三维空间）显现着怎样的投射规律，如何描述和表现四维世界的画法几何，便是杜尚所关心和想要探讨的内容，也是杜尚制作大玻璃的目的所在。尽管大玻璃这项研究仅进行了一半，未能最终完成，但是这种关心四维世界如何投射的问题本身却有着深刻的价值和意义。因为自文艺复兴以来，人们的视点都仅仅是关注平面中的三维物体，仅仅只关注三维空间中的对象本身，忘记了维度投射的因果顺序，忘记了透视的真正的含义。面对这些问题，重新关注空间投射器，重新对其研究，重新对其理解，重新试图用这种透视的装置追溯探讨四维世界的投射关系本身便是杜尚所做的以"新娘"为题的大玻璃窗的真正意义所在。从这个层次上来看，我们可以说：大玻璃是杜尚进行研究四维投射作图法和三维画法几何的投射装置。

（三）两种投射

上面我们所说的丢勒的投射和杜尚的投射，实际在本质上还存在着明显的区别。首先：丢勒是将三维世界在二维平面的投射作为研究对象和研究目的。而杜尚所关注的是四维在三维中的投射问题，并力图在三维世界中找到四维的投射法则。尽管两块玻璃所形成的"框"与"窗"是相似的，但在观念和本质上却是根本不同的。这两种不同的投射，一种是思考外在对象本身在玻璃上的投射，一种是内在观念的意识本身在玻璃上的投射，二者实际上是反映了自文艺复兴以来的空间观念与现代空间观念在本质上的不同。

projection as the basis of understanding three-dimensional worlds limit our understanding and expanding on spatial dimensions.

(B) Duchamp's glass

With progress of science, foregoing understanding and expression of space were re-examined in the 20th century. Painter Duchamp made a large glass named "bride" (similar to Dürer's window) to create a new way of perspective and explore a projection theorem which can project four-dimensional world to three-dimensional one.

Just like three-dimensional objects can be projected onto two-dimensional plane, the three-dimensional world is the result of the rejection of four-dimensional world! In this way, images projected on Duchamp's glasses are the three-dimensional world itself. Duchamp focused on projection laws and descriptive geometry for four-dimensional world. Although his study on glass failed halfway, but it still is full of profound value and meaning. Since the Renaissance, all the people stressed on three-dimensional objects in the plane and the object itself, disregarded the causal sequence of projection and the true meaning of perspective. Duchamp's glasses "bride" focused on research and understanding of space projector, reattempting to discuss relationship of four-dimensional projection itself. This is the real significance of the glasses. From this aspect, Duchamp's glasses are projector for four-dimensional projection graphing and three-dimensional descriptive geometry.

(C) Two kinds of projections

Projections of Dürer's and Duchamp's are different in nature although they are similar in form of glass and frame. Dürer wanted to study projection from three-dimensional world to two-dimensional world, which is external object projection on the glass, while Duchamp focused on four-dimensional projection in three dimensions, and tried to find the projection theorems. Differences between the two projections reflect the differences of space concept between Renaissance and modern times.

2. Experience space projection and conscious space projection

Projections mentioned above remind me two projections I have experienced in different levels and significances. One is the settlements

二、经验空间的投射与意识空间的投射

上述丢勒和杜尚的两种不同层次与意义上的投射，令我联想到我所经历的两种类似的不同层次和意义上的投射。一种是我所进行的聚落研究的过程中，为记录聚落空间所进行的聚落平面图测绘时所感觉到的投射，另外一种是我在进行设计时，我头脑中的意识空间在现实三维空间中的投射。

（一）我测绘的聚落平面图

在过去的十几年中，我曾经对于聚落进行研究，而其中对于聚落进行测绘是必需的工作。在聚落的调查过程中，我曾对处于三维状态的聚落进行过总体空间关系的测绘和记录。这个测绘和记录工作是通过对聚落平面关系的测量来完成的。在记录过程中，原则上我是将聚落中的住宅、树木、家畜房屋等眼睛所能看到的所有东西都记录在一张平面的纸上。其结果在平面图上所记录的图纸内容本身就成为聚落空间关系在二维平面上的投射。事实上，在从聚落空间向聚落总平面图投射和转换的过程中，实际上我只记录了三个相关的"量"的关系。一是聚落中存在的住宅的方向（各住宅的朝向），二是住宅的大小，三是住宅之间的距离。由于对这三个数学量关系的记录，从而完成了聚落总体关系图的绘制。这个总体平面图并不是丢勒透视学意义上的平面图，而是一个空间构成关系的平面图。它不是一个有关三维场景的记录，而是一个空间构成关系的记录。而这个空间关系就是三维空间在二维平面上的投射。由此，二维的聚落平面图与三维的聚落空间之间也就存在着投射和像之间的对应关系。值得注意的是，总体聚落平面图中所记录的，实际上是一个数字和量的关系，也正因为如此，聚落的总体平面图又可以用如下的公式来进行表示：

即：聚落的总体平面图＝住宅的面积＋住宅的方向＋住宅间的距离

另外一点值得注意的是：在我一个个地将聚落的空间投射到聚落总平面图上的过程中，实际上我的身体也参与了一遍聚落整体的建造过程，具体来说就是我所进行的聚落总体平面图的测绘工作的过程，实际上就是聚落建造者建造聚落过程的镜像存在。也就是说，聚落的建造者在建造聚落时同样地经历了一个

space mapping projection in my research on settlements, the other is projection from my conscious space to real three-dimensional space during my design process.

(A) The map of the settlement plan

In the past ten years, I conducted research of the settlement. In the process, mapping is a necessary work. I mapped and record spatial relationship of three-dimensional settlement by measuring the plan. In principle, I record residential settlements, trees, livestock houses and everything can be seen on a paper. The results actually are projections of settlement spatial relationships in two-dimensional plane. Actually, I record three related factors of "quantity": the direction of the settlements, the size of the houses and the distance between houses. It isn't perspective of Dürer's projection, but a plan reflects spatial relationship. It is not a record of three-dimensional scene, but a record of space constitution which is projection of three-dimensional space to the plan. In this way, relationship between two-dimensional plan and three-dimensional settlement space are set up, just like relationship between projection and its image. It is noteworthy that the record actually is relations of digits and quantities. Thus the plan of the settlement could be expressed by the formula as follows:

the overall plan of the settlement = residential area + residential direction + distance between residences

Another point worth noting is that my body involved in the overall construction process of the settlement during the process of mapping which is the image of construction process. Construction activities is a process of converting residential direction, the distance between the residential and the residential area into three-dimensional space. We can say that the drawing of the total settlement plan equal to the design process of settlement in this level.

将住宅方向、住宅间距离以及住宅面积投射和转换到三维空间中的过程。在这个层次上我们可以说：聚落总平面图的绘制过程，等同于聚落的设计过程。因此，聚落建造者的意识空间也可以通过以下三个量的关系来加以表现和诠释。

即：意识空间图=村落总平面图=住宅的面积+住宅的方向+住宅间的距离

在这里必须指出的是：对于聚落的调查过程，实际上又是聚落空间投射和转化为我的意识空间的过程。由于投射到我意识中的空间同建造聚落的人们数量化状态的意识空间之间形成了一个相互的镜像关系，因此在我的聚落调查过程中，从聚落的实际空间转化到头脑中的数量空间的过程，也正是他们从数量状态的空间转换到实际空间中的过程。因为对于聚落的建造者们来说，意识空间的本身是一个数量状态的存在，而聚落空间的本身是这种数量空间状态的客观投射的结果。因此在我调查聚落时，从聚落空间组成中所感受到的意识空间实际上是聚落建造者们的集体意识的集合，并且这个感受过程同时也是一个从聚落的空间组成感觉聚落居住者的意识空间的过程。

（二）意识空间的投射

上面我们所说的聚落的空间组成实际上是聚落的建造者意识空间的投射结果的事实，说明了以下几个问题：

如果说：聚落空间是以三维空间的形态存在的是事实（这个事实是不可否认的），并且聚落空间又是数量化的意识空间投射的结果，那么人的意识空间本身就一定是四维空间的存在。

如果说：建筑师所设计的建筑空间是一个三维空间的存在，那么它一定也应该是建筑师头脑中的意识空间在三维空间中投射的结果和事实。

如果说：建筑师所设计的建筑空间是建筑师头脑中的意识空间的投射的结果，那么建筑师头脑中的意识空间一定是一个四维状态的存在。

如果说：聚落的建造者头脑中的意识空间是一个数量状态的存在，那么建筑师头脑中的意识空间也应该是一个数量状态的存在。

如果说：意识空间是一个数量状态空间的存在，那么四维空间向三维空间投射时的画法几何和投射媒介只能是数与数学、几何学的存在。因此在设计的

Therefore, conscious space can be understood by the following three factors: conscious space map = the overall plan of the settlement = residential area + residential direction + distance between residences

It is the the process of projection and transformation of the settlement space to my awareness space that the investigation process of the settlement actually is. Space projected to my consciousness is the image of awareness space of builders' state, so the translation from settlement space to awareness space in my mind is equal to the converse translation. Conscious space is a digital existence for builders and settlement is the result of projection from the digital space. I think settlement space is collective consciousness of builders and my survey experience is also a process forming dwellers' conscious space from settlement space.

(B) Projection of awareness space

All the things we discussed are actually results of conscious space projection of builders, this fact tell us:

If we believe settlement space exists in the form of three-dimensional space (this fact is undeniable) and it is a projection of conscious space, then people's conscious space can be a four-dimensional space.

If we believe space is a three-dimensional existence, then it must be results of projection from conscious space to three-dimensional space;

If we believe space is the results of consciousness projection from architects then the conscious space of architects should be in a four-dimensional state.

If we believe conscious space of builders is in a digital state, then conscious space of architects is on the same page.

If we believe conscious space is a existence of a digital state then the descriptive geometry and projection media for projection from four-dimensional space to three-dimensional space can only be a mathematical geometry

过程中，我的意识空间通过我的设计投射为三维的建筑空间。在此有两点应该指出：

(1) 我的头脑中的四维状态的意识空间源于我对空间的经验。

(2) 我的设计过程不过是通过几何学的手段将我头脑中呈数量状态的意识空间投射和转化到三维的空间。尽管我头脑中的意识空间用眼睛看不到，这似乎如同二维平面中的生物是不可能看到三维空间物体的存在是一样的。但我的意识空间一旦经由投射成为显现物，转换为三维的空间对象，我的头脑中的意识空间就会由于投射而成为客体得以表现。

我以为：设计者对于空间的设计，实际上是设计者意识空间的流出和显现，是意识空间的投射和意识空间在维度上的转化，是意识空间在三维世界投射结果的体现。而设计师所设计的空间本身正是设计师本人的意识空间的空间投射器。如果我们能够从空间维度的转换和投射的角度来认识和理解建筑的话，那么我们就可以说：

建筑是四维状态的意识空间在三维世界中的投射！

（三）建筑是意识空间的展现

上述"建筑是四维状态的意识空间在三维世界中的投射"的理解实际上是我们摒弃"视网膜"的建筑走向意识的建筑的开始。在这里需要强调的是：所谓的我们的眼睛"看"和"看到"是两个完全不同的概念。在我看来眼睛就像是一个屏幕，它具有两个面，一个是现实投射的面，另一个是意识投射的面。当我们"看"时，总是希望视网膜不断地受到刺激，看到物，而这个时候你不会产生梦，你的意识当中的"风景"也不会展现。因为当现实的东西不断地被投射到视网膜上，也往往是外界的刺激集中灌输进大脑的时候，所以我们才说睁眼看世界。但当我们闭上眼睛睡觉的时候，我们会做梦，因为在我的理解中这是大脑里的东西投射到视网膜上的缘故。而这个时候，你就会看到大脑里的那个世界。因此，如果从这个意义上来理解，我倒是希望这种意识中的风景能够得到展示，因为这也是一种生活状态。从"意识的风景到现实的风景"反过来通过"现实的风景唤起意识中的风景"：或许能够成为我们对于建筑的另外一种理解以及思考的状态和出发点。

existence. Therefore conscious space can be projected to three-dimensional space in my design. There are two things need to be brought up:

1) Conscious space of four-dimensional state comes from my experience.

2) The design process is the way I project conscious space to a three-dimensional space. Just like objects in two-dimensional plan can't see objects in three-dimensional world, we can't see conscious space in mind. But projecting to three-dimensional world, awareness space can be converted to objects.

I believe the design process is actually a translation from conscious space to three-dimensional space. Space designed by architects is projector of their awareness space. To understand architecture from the aspect of spatial conversion we can say:

Architecture is the projection of four-dimensional state space in three-dimensional world!

(C) Buildings show conscious space

If we believe "Architecture is the projection of four-dimensional state space in three-dimensional world", we just abdicate visual buildings and begin to search conscious ones. "To watch" is different from "seen". Human eyes like a screen have two faces: one is for realistic projection; another is for conscious projection. When watching something, we get objects visually but we can't get dream and "scene" in our consciousness. It's because when objects are projected onto our retina, external stimuli are getting into the brain. When eyes closed, we have dreams. In my opinion, objects in our brain are projected onto the retina and you can see the world in your brain. I'd hope that this state of consciousness can be displayed because it's also a state of life. Arousing realistic scenery through conscious scenery maybe a new state and the starting point for our understanding of architecture.

王昀 设计作品

Works by WangYun

白色方体空间设计操作过程中的思考

文/王昀

对于建筑而言，实际上一直存在两种论点：一种是将建筑视为"容器"理解，另外是将建筑作为一种"造型"来看待。从某种意义上讲，将建筑视为"容器"的观点，似乎还在关乎建筑的空间问题，但是造型论则无疑地将建筑与雕塑相提并论。仔细地想来，实际上如果仅仅将容器的空间作为一种造型来理解，确切地说将空间作为一种纯粹的物理性的对象物来进行追求和表述，实际上这种将建筑视为"容器"的观点本身也还是一种造型论的立场。这两者本身仍然是一种将建筑仅仅作为物质表达的物质论观点。

对于建筑的物质性存在的事实，尽管不太会有人来对其加以否定，但是尽管结果均表现为物质，但由于各自的出发点的不同，带来的结果实际上也不同。这一点从对于聚落的样本观察中会看得非常清楚：实际上作为没有建筑师的建筑——那些聚落中的建筑，是那些非建筑师的居民为自己生活的需求而建造的建筑。作为结果，其产生的形式与那些为了建造一种形式而进行建造的活动本身表面上看是没有太大区别的，但是如果经过体验，就会发现：前者的建筑是一种生命体的存在，关乎建造者的意识与经验；而后者则如同是一个标本和蜡像的存在，是一个没有生命的尸体或偶像。

我们注意到：聚落中的那些没有建筑师的建筑是聚落中的人们根据自身的经验进行的自我需求的表述，而作为职业的建筑师往往经常性地陷入为了别人，或为了某种目的而去帮助别人去表达的境地。这就如同前面所比喻的：生命体本身是生命体自身的表述，而蜡像则是试图去表达生命体本身。

我们都知道，建筑的设计实际上是设计一种生活，而这种设计生活的过程有赖于一个基本的判断，就是建筑的经验性判断。这种经验性判断的根源实际上是作为建筑师如何进行设计和操作问题。如同聚落中的没有建筑师设计的建筑一样，居民们的建造过程是依赖于他们自身的意识存在的问题，是让设计本身与设计者本人的意识之间产生互动和关联，是设计的根本状态，是对于设计本源的还原。事实上，经验的产生同样地是与人的身体本身密切相关联的。而人的身体问题实际上存在两个部分：一个是作为肉体的身体，而另外一个是作为一种意识存在的身体。作为意识存在的身体的本身事实上又含有两个部分：一个是作为视觉本身存在的，另外一个是包含意识，确切地说是包含记忆本身而存在的。这件事情实际上变得复杂，如果说记忆与经验有关，那所谓的经验本身，实际上还是还原成为记忆，而这种记忆的存在，在意识之中的低层，往往又通过活动将其"唤起"。而"唤起"这个概念，在建筑的操作过程中起到的是一种具有结果性意义的东西和状态。我们通过设计来进行具有操作性的工作同时，我们的经验和记忆被"唤起"并通过手透射到所设计的空间之中。而当人活动在被投射有内容的空间中时，由于投射的对象物中存在着透射过的信息要素，这种要素与活动在其中的人之间发生信息沟通上的关联性，信息要素本身便会唤起活动在其中的人意识中的相关联的要素和空间内容。而这样的一种空间操作，其结果尽管同样地表现为一个物质空间结果，但本质上并不是局限于建筑的造型问题，更不是单纯的空间比例与造型的关系问题，而是在空间中注入生命的过程。这种"唤起"的设计操作性的存在是设计的出发点，它实际上与设计师的记忆和经验直接发生着关联，是设计师的记忆部分不自觉地投射到空间之中的过程，是拥有用这种被透射了记忆的空间同时去唤起行走于空间中的人的记忆的设计过程，更是独立于"容器论"和"造型论"之外的对于建筑的第三种设计的理解和设计方法。

"唤起"这个感念在这个过程中所扮演的角色实际上是与建筑的结果以及目的和意义密切关联。"唤起"概念本身，事实上隐含了一个问题，那就是唤起什么的问题，究竟唤起的东西作用于使用者的是什么？"给人以感受"，这只是一种常用的语言，难道不经过空间操作的东西就不给人以感受么？显然不是，我们这里之所以不用感受，而是采用"唤起"这个概念，是因为感受这个概念的本身只是一个知觉在肉体上的反应的概念，是一种表皮性特征的存在。而这里所谈的"唤起"，具有将埋藏的东西加以再度重现的含义，有一种从表皮的感受触及意识层面深度的驱使性。实际上更大的期待是寄希望能够在大脑中产生一种情怀，这种情怀产生的同时，能够直接地在人的大脑上产生一种景象，而这种景象能直接地指向对于记忆的勾起，而勾起的东西，往往是所谓的"触景生情"，这种"触景生情"的作为在我看来恰恰是建筑师在建筑操作的过程中的一个非常重要的工作，也是我们设计追求的所在。

面对这样的思考，我们会面对另一个问题，那就是如何让人能够触景生情，我们的这个"景"应该如何加以操作和完成？在我看来，这个"景"的操作问题，最大的问题和难点就是如何让更多的人，在这样的空间中都能够被勾起些什么，能够唤起些什么。

我们不得不承认，当前的世界，信息的多样性问题已经不是过去的封闭时代的单一性表述了。具体而言，由于交通便利、电脑、移动电话、微博的存在，我们当今的时代中，距离的概念已经在发生着质的变化。面对物理的、真实距离存在的错乱和丧失，国家与民族之间文化距离感的逐渐消失，对于建筑师而言，由于其所建造的建筑的未来使用者的多样性和公共性的存在，采用某种具象的操作只会对于那些仅仅曾经存在过与设计者之间具有某种相同具象场景相关联的人才能够产生相同的感受。但是如果采用这样的一种具象的操作，往往也只会让人的想象力因视觉的满足，或具象的对象物本身具有了明确的意味指向，使得使用者的意识因为获得了具体的信息回复而使意识中"唤起"的操作功能关闭和停止。

如果我们不是采用上述进行的一个具象的操作，实际上我们可以采用将设计师的记忆加以呈露和展开的方式来进行空间的布局与展示。由于人的记忆不是照相机（除非我们在进行设计的操作时将照片放在眼前进行摹画），记忆中所呈现的风景含有某种不具体性。即使是记忆中存在着某种模糊中的具体，那个像实际上也是不确定的。然而建筑的表达必须是确定的，因为建筑本身一定要用具体的形态加以表述，一定要以具体的色彩加以表达，而实际上当这两者产生矛盾的时候，采用一种将界限模糊，将边缘模糊，将造型模糊、将意向模糊的操作方式。具体地，我们将建筑形态采用抽象要素来表达，将空间的内部进行最基本的操作，去除空间的各个组成面中的具有具体指向性要素的存在，同时对建筑周边具象的风景进行选择性的操作。通过这样的过程，使得平时大脑希望获得具象信息的习惯性操作的使用者在建筑空间里进行移动的过程中，当整个身体被放置在一个不确定的、没有像的环境中，其眼前的景象系统被强行关闭，进而将其头脑中的景象唤起。当然，空间的操作对于"唤起"的操作是非常重要的，而这种操作的基础同样来源于设计师记忆本身的工作。整个空间的铺述和展开在整个的操作上是重要的。

明确了这种对于建筑的基本理解，所谓的白色方体空间的操作问题则不言自明。因为方体本身有一个可以操作的空间界面，这种界面是一个领域的存在，同时也是一个抽象的存在。关于白色，在我看来，白色存在抹煞了所有存在于人的意识之中的具有惯性的具象请求，它可以让具象的对象消失，从而能够更大程度地邀请和唤起存在于人的大脑中的潜像的浮出。

The Thinking during the Design of White Cube Space
By Wang Yun

Talking about architecture, there are two arguments which have been discussed: one is seeing architecture as "container", and another is considering it as "shape". Speaking from some kind of sense, it seems the "container theory" concerns with the "space", otherwise the "Shape theory" undoubtedly puts architecture on a par with sculpture. If the space of container is seen as a shape, to be more exactly, formulated and explored space as a purely physical object, the perspective of considering architecture as container is still limited in "shape theory". Both of the two views are still seeing architecture as the expression of substance.

Though little would negate the fact that architecture exists as substance, the problem is although the result appears to be a substancial one, based on different method, Distinguishing consequences come out. The observation of the settlements suggests clearly: in these settlements, the buildings which are designed by non-architects are carried out by the requirements of inhabitants' daily life. Seemingly the form of these buildings is not so different from those which are originated from a certain shape pursuits of the architects. But If we analyze the phenomena with experiences, we'll get a conclusion that the previous one is a living existence of life, which is related to the consciousness and experiences of constructors, and relatively the latter one is more or less kind of specimen or wax statue, a lifeless dead body or idol.

We have been aware of that: the architectures without architects, which are the expression of self-requirements, are built according to the habitants' life experience, while the professional architects are always trapped into a circle of speaking for others or expressing with a certain purpose. That is just the analogy given earlier, the living body itself is the expression of its own life, while the wax statue which is trying to express life.

As we all know, architecture design is essentially a life design. This kind of life-design depends on a fundamental judgment which refers to the architectural experiential judgment. The source of the judgment is how to design and manipulate as an architect. As the phenomena of architectures without architects, the inhabitants' construction is the existence relying on their own consciousness, the interaction and correlation between the awareness of design and the designer himself, the fundamental state of the design and the return of the design origin. In fact, the generating of experience is closely associated with human body. The human body can be understood as two parts, one is the physical existence, and the other is conscious one. Again, the conscious human body can be divided into two parts, one is existed as vision itself, while the other existence contains consciousness, or speaking exactly, contains memory. The matter goes more complicated. If memory is considered relating to experience, the experience itself is the return of memory.

The existence of memory is always on the bottom of the consciousness, which could be aroused by the activities. During the manipulating of architecture, the concept of "evoking" plays a significant role. In the process of design, experiences and memories are evoked and reflected on the space designing. When the users use the space which reflects memories and experiences, due to the space containing much information elements, the information would communicate with the users and actively associate elements and spaces which are kept in the users' memories. Although such space manipulating results to a physical space, it's not limited in the architectural shaping, more than simple problem of proportion and form. Instead, this method injects life into the space. Taking "evoking" as the start point of the design actually associate with the memories and experiences of designers. It suggests a process that the memories of designers unconsciously reflect on the space and the procedure that utilizing the space contains much information to evoke the memories of users. It's a way of designing independently from "container theory" and "shape theory".

In fact, "evoking" is a special role of the whole design procedure, which relates to the results, purposes and significance of architecture. The concept implies a problem for itself, what should be evoked, what should be given to users. "Touching the feel" is just a common word. Someone would doubt, doesn't the space without "evoking" design touch the users? Apparently not. We use "evoking" instead of "feeling" here, in order to explain that feeling is only a perception of physical reaction and superficial presence. The "evoking" I talk about means reproducing the covered thing and reaching the conscious depth from epidermal feel. We expect more that we can evoke a special emotion, as the emotion were evoked, a scene would be emerged directly, and then the scene call the memory deep in mind. All of these are so-called "a scene which recalls past memories". In my opinion, it is a very important job of the architects in the process of architecture design, also the pursuit of design.

We have to acknowledge that in the modern world, the multiple issue of information could no longer be described by unity as in the past of close time. Specifically speaking, as a result of the convenient transportation, and the existence of computer, the mobile phone, and micro blog, in our now times, the feeling of distance was changing substantially. Facing confused and loses of the physical real distance, cultural sense of distance which betweens country and nation is disappearing gradually. For architects, because of the multiplicity and publicity of the future users of the construction, if architects use some representational operation, it can only make someone that has similar representational related scene to have the same feelings. If we use such kind of representational operation, as it will only makes a person feel vision satisfaction, or representational object itself has the explicit meaning, it will cause the user's consciousness "arousing" the operating function close and stop.

Thus instead of using representational operation as stated above, we may reveal and launch designers' memories in the way of carrying on the layout and exhibition of space. Because people's memory is not camera (only if we draw picture in front of us when carrying on operation of design), the scenery which presents in the memory is some kind of unspecific. Even if the memory has some kind of fuzzy concrete, that image in fact is also indefinite. However, the expression of architecture must be definite, because the construction itself must be expressed with concrete shapes and concrete colors. But in fact, when these two have contradictions, we use one kind of operation to make the boundary, the edge, the model and the intention fuzzy. Specifically, we will use the abstract essential factor to express the form of architecture, carry on the most basic operation in the spatial interior, remove the specific directional factor, and carry on the selective operation to concrete scenery around the architecture. Through such process, when users who are getting used to the habit of obtaining concrete information walking in the architecture space, his entire body is laid aside in indefinite environment without concrete factors. Thus picture system of his body was closed down forcefully, and then the pictures in his brains arouse. Certainly, the operation of space means a lot to the arousing operations, but the foundation of this kind of operation originates similarly from designer's memory work. The entire space's statement and expression are important in the entire operation.

Has been clear about the basic understanding of architecture, the so-called white cube space's operation problem is self-evident. Because the cube itself has a spatial surface that could be operated, this kind of surface is an existence of domain, and is also an abstract existence. About the white, in my opinion, the white existence obliterated all existing inertia of concrete requests in human's consciousness, thus it may let the concrete object disappear, invite and arouse emersion of latent images existing in human brain in a greater degree.

善美办公楼门厅增建 北京
Annex Foyer of Shanmei Office Building, Beijing, 2002

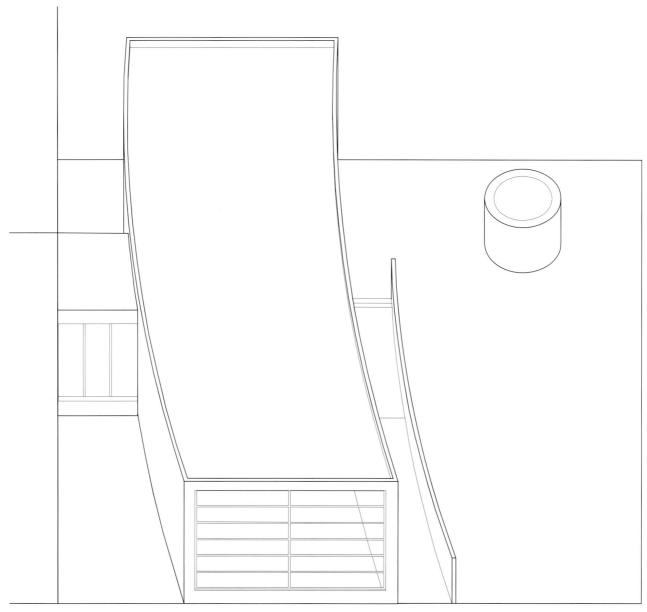

轴测图 / Axonometric drawing

这是一个在原有办公楼东侧加建一个门厅的小项目。原有办公楼的入口在建筑的南侧，业主希望将主入口改设在办公楼的东侧。由于现状楼的东侧只有一个设置在走廊尽端的门，因此在其延长线上增建怎样的门厅便成为设计的关键。

增建门厅基地的南北两侧有两栋建造于不同时期的住宅楼，南侧的为红色，北侧的为米黄色，原有场地内还有几个零散的小仓库，一条狭长小道使基地与城市道路相连，场地整体呈灰色调，气氛凌乱。设计时，我们试图在这个复杂的场地中间放置一个醒目的对象物，同时让这个对象物本身成为这个凌乱环境场所中的视觉焦点，从而也突出了入口的显著位置。为了更加强调入口的主体特征，我们将原有米黄色的办公楼施以黑色，使其淡化，并与周围的住宅楼共同成为围合办公楼前广场空间的背景要素。增建的门厅采用单纯的白色几何形体，运用摆放一个微弯曲方筒体块的方法使之成为小广场上的视觉中心。

增建的门厅前面，设计时布置了一个小广场。小广场的铺砌处理采用黑灰色的鹅卵石，与现场既存的混凝土方砖地面产生对比。同时在既存方砖和鹅卵石铺砌之间用150毫米宽的混凝土围合了一个如同画框般的范围，用以在平面上更加强调和突出入口的特征。广场中的黑灰色的鹅卵石是从北京近郊平谷区的河谷中捡来的，每一个黑灰色的鹅卵石的椭圆直径大约在8毫米左右。通过这样的一系列动作，在场地中重新注入黑、白、灰的色调并恢复场地内部自然与舒展的节奏感。

在建筑完成的过程中，现实的阻力是巨大的，比如设计中入口的大门，其实际的尺寸是2.0米宽和2.7米高，这样的门的尺寸被做门的厂家认为是超出了他们所能够制作的常规尺寸，所以他们一直建议将大门改为一个双开的子母门。理由是这个大门的尺寸过大会使未来的成品超重不能开启从而影响使用。这样的一个小问题曾一度成为业主、厂商与设计师之间的矛盾焦点。事实上，当门安装完成后，大门却能够轻松推开，而超出原有的种种不利的设想。

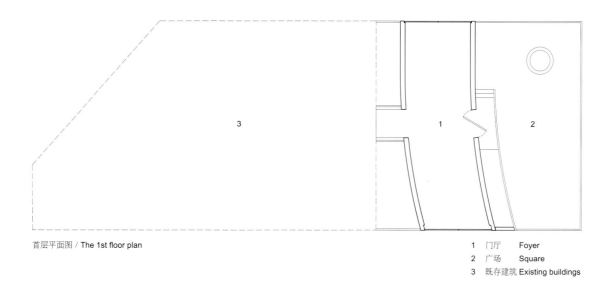

首层平面图 / The 1st floor plan

1 门厅 Foyer
2 广场 Square
3 既存建筑 Existing buildings

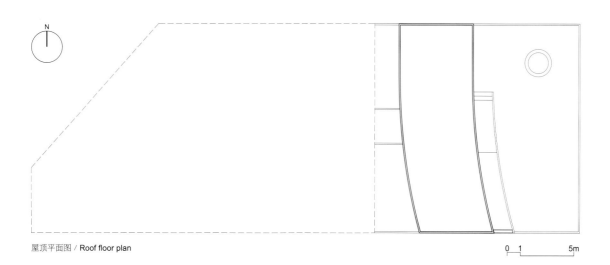

屋顶平面图 / Roof floor plan

0 1   5m

This small project is adding a foyer on the east of the existing office building whose original entrance is on the south. The owners hope to re-establish the main entrance on the east side of the building. What kind of entrance is suitable for the building on the extension line of corridor is crucial for the designer because there is only one door at the end of corridor. On the site, there are two residential buildings of different times (the south one is red and the north one is beige), a few warehouses and a narrow channel linking the site to the city. The whole site looks grey and in a mess. Thus we tried to insert a visual focus in the place and stress the role of entrance. To emphasize the characteristics of the entrance, we painted the original beige building black and made it the background of the plaza space together with the surrounding residential building. The annex foyer is a pure white geometry and become the visual center of the small square by placing a micro-bending cylinder block in it.

We designed a small square in front of the foyer, paving black and gray pebbles there to contrast with existed concrete bricks. A concrete frame (150mm in width) was inserted between bricks and pebbles to highlight the entrance. The black and gray pebbles were picked up from the valley in Beijing Pinggu District, and the oval diameter of each pebble is about 8mm. These series of actions bring black, white and gray tone and nature rhythm to the site.

We met huge difficulties in completion of the design. For example, the size of the door (2.0 meters wide and 2.7 meters high) is considered to be too big by the manufacturers who recommended changing it by a double door in case that the heavy door might don't work efficiently. Such problem once became the focus of owners, manufacturers and designers. In fact, non adverses happened after the installation of the door.

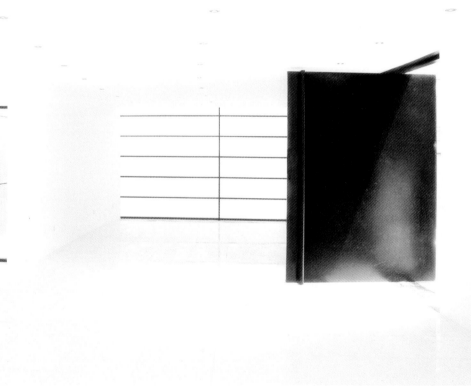

60平方米极小城市 北京
"60 Square Meters" Minimum City, Beijing, 2002

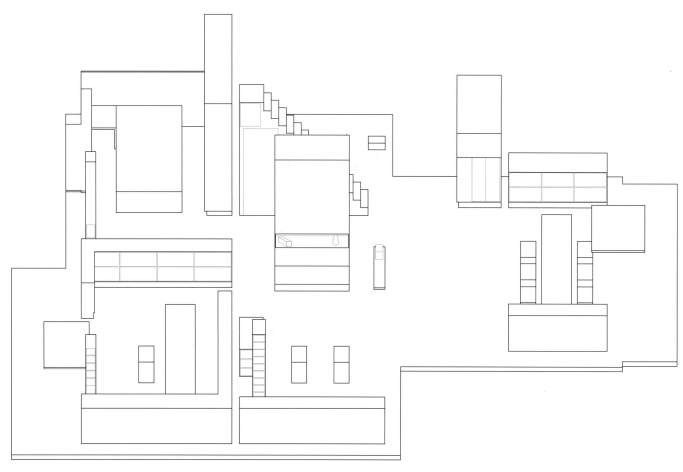

轴测图 / Axonometric drawing

2001年来北京大学任教时，学校按规定租借给我一个三室一厅的住宅，条件是租期5年。这是一个整体为6层的砖混结构住宅楼，一梯两户，居室内部为三室一厅的布局，南侧有两间卧室，北侧有一间，入口的对面是厨房，厨房的南侧是客厅。

北京的城市尺度在不断地扩大，街道被一而再再而三地拓宽。过去的人从街道上回家的过程是从宽马路→胡同→小巷→入院→进入家，这个过程中，城市整体尺度的递减与人的心理尺度的递减过程是相互吻合的。反过来，从家里出来到城市大街道的过程也是伴随着人的心理尺度递增而存在的。现实中的城市空间这种心理的过程是紊乱的。从家里出来，经楼梯三绕两绕，打开楼门是两座住宅楼之间的空地，人在"空地"般尺度的街道上行走。人的心理也伴随着空间忽强忽弱的变化而变化。以往的逐渐变化的心境在当下成为"大起大落"的状态。

实际上这个租借来的住宅现状与当下城市的凌乱状况也非常地相似。没有改造之前，住宅内部的空间之间没有任何的"渐变"与"过渡"。其中卫生间的门被设计成直面客厅是最不能忍受的。这个住宅的尺度不是很大，使用部分只有60平方米，由于墙体均为承重结构，所以无法进行根本性的调整。

面对这些问题，在进行新的改造时力图在自己居住的内部空间中将符合人的心理空间的渐变的尺度关系加以还原。尽管房子的物理空间很小，但实际上其构造关系与城市是相互对应的。如起居室与城市的广场相对应、餐厅与城市的餐厅对应、卧室与城市的旅馆对应、书房则与图书馆等公共场所相对应。而将所有这些功能之间加以串并联则是街道的功能所在。考虑到这些，于是，一个与城市构造相对应的风景便可随之展开了。家的街道可以很窄，仅600毫米。这样窄的街道纵深确是可观的。如果加上餐厅和卧室之内的"街道"有10米。希腊米科诺斯岛上将厕所放在住宅楼梯下方的智慧也被投射到新的改造之中。由于家里的楼梯是一个壁式的柜子，厕所门与柜子门合为一体，卫生间的门成为壁柜的一部分，从而将厕所隐藏。"街道"的另一侧是一个作为"橱窗"的壁柜，门采用推拉方式，伴随门的开启与闭合，"广场"和"街道"之间的空间性质与关系便能随着其变化而改变。

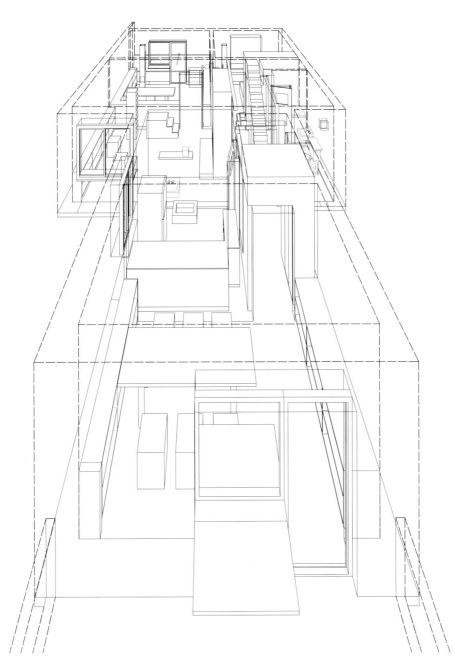

极小城市鸟瞰图
Bird's eye view of tiny city

When I entered Peking University as a teacher in 2001, the university rent me an apartment with 3 bedrooms and 1 livingroom. 2 apartments share one stair in the six-story residential building. The apartment has two south rooms and a north one. Opposite the entrance is the kitchen, and living room is on south.

Urban scale in Beijing is constantly expanded, and streets become wider and wider. Before, people experienced the process of "road→lane→alley→courtyard→house" to get home. In the process, the urban scale and the human psychology scale decrease simultaneously, vers versa. But now the psychology process has been disturbed: coming out of apartment, we go downstairs and reach the open space between residential buildings, and then walk on the streets of "open space" scale. The apartment shared the similar situation with the messy urban state before renovation. There's no transformation spaces. It's intolerable that the bathroom door was facing the living room. The apartment is only 60 m^2. All walls are load-bearing made it impossible for fundamental adjustment.

In the renovation, I tried to restore scale relationship to meet psychological space demands. Although physical space of house is small, it corresponds to spaces of the city: living room equals to square, dining room means restaurant, bedrooms stand for hotels and study like public spaces as library. The function of street is to link all these spaces. Therefore a city likely landscape established. Streets in house could be very narrow (only 600mm) but quite impressive in depth, it can reach 10 meters added with length in "restaurant" and "bedroom". I learned from example in the island of Mykonos, Greece (toilet is put below stairs in the house there) and hide the toilet in the cabinet (stair in the apartment is in the form of cabinet fixed on the wall). Another cabinet with sliding door is on the other side, with the opening and closing of the door; space nature changes between "square" and "street".

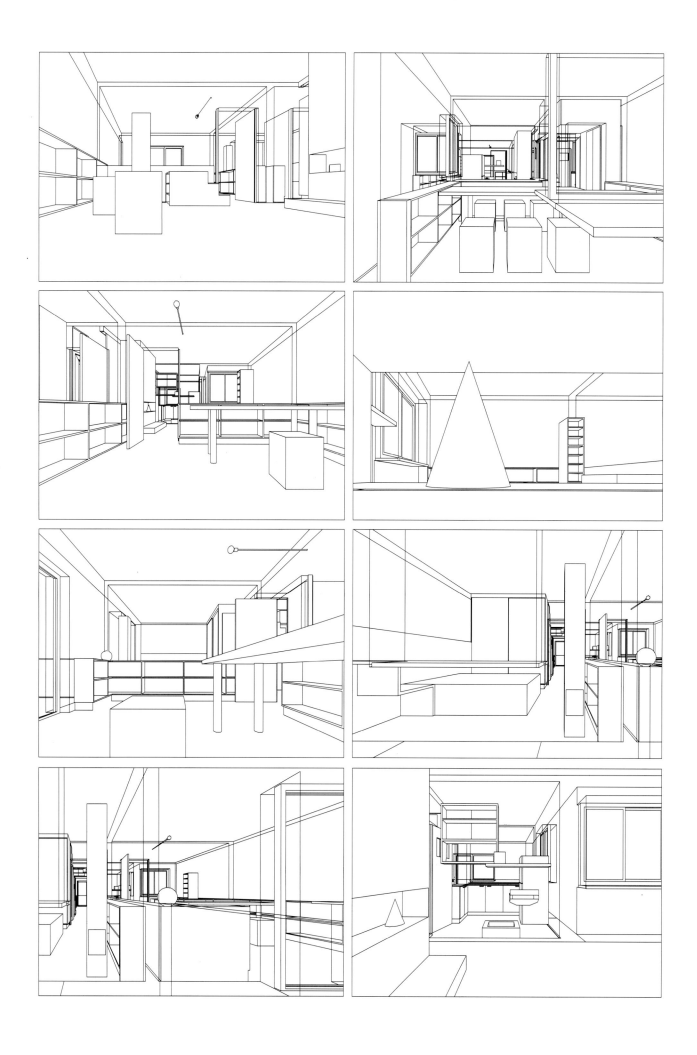

庐师山庄A+B住宅 北京
The Lushi Hill A+B House, Beijing, 2003

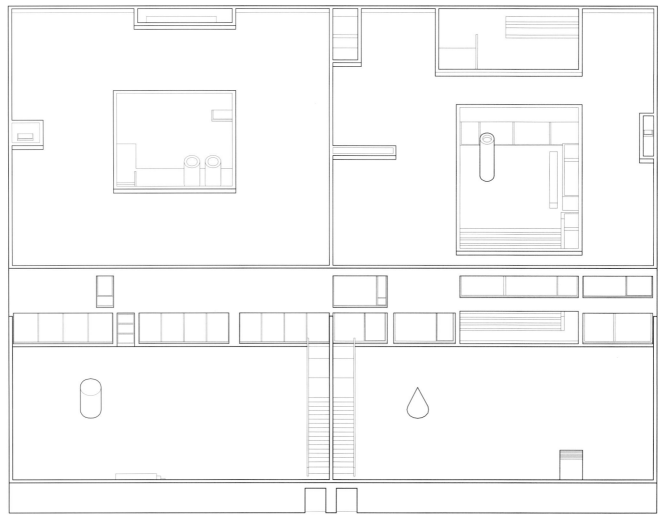

轴测图 / Axonometric drawing

白色是一个覆盖的色彩，包容丰富。白色是一个合成的色彩，由7色混合而成。白色是一个消失的色彩，即所谓"实景清"后而"空景现"。在白色空间中消失着确切的尺度和距离感，而白色界面所产生的消失感，还原着设计者纯粹的观念和对空间场景的梦幻。伴随着颠倒和漂浮的无距离感空间的体验，体验者最终可以获得"无画处皆成妙境"的空间经验。

住宅A+B是两个联立式的小住宅，位于北京西山八大处附近的庐师山庄别墅区中。该别墅区是由52栋小住宅所构成，住宅A+B是其中的两栋，由于建筑内部空间表现出"刚"与"柔"的性格，所以这两栋住宅又被称为"王"与"后"。

住宅A+B是山庄中最大的住宅。山庄中普通住宅的面积约为300～400平方米不等，这两幢房子的地上和地下面积加起来约700平方米，且在住宅的东侧两栋分别各延伸出两个大小为18米宽、进深为12米的内院，内院中有分别有楼梯能够直通到住宅的地下室部分。两栋住宅由两个长和宽均为18米、高为7米的方盒子拼合联立在起，并且均为地上二层地下一层。两个住宅的空间性格，一个刚性，一个相对暧昧柔性。尽管房子本身抽象，但却希望能够给使用者带来更多的可能。未来生活在其中的每个人的生活经历是不同的，想象力也不尽然相同，而别墅本身应该给居住者提供更多的想象和可能性的空间。

这两个住宅在整体的设计上采用抽象的白色箱体进行空间构造。在内部空间的组织上，设计者力图将其意识中的风景物象化。室内穿插游走的散步路径，步移景异的空间景，用抽象的景观作用于体验者。眼前幕幕抽象风景的呈现，在激唤起体验者自身经验的记忆与联想的同时，使体验者获得高次元意识的直观。

这样的两个看似简单的建筑在完成的过程中却费尽周折，首先面对的是施工质量的要求问题：如何在实际施工过程中认真地做到"横平竖直"，是一个不小的难点。在与工人的交流过程中，工人们坦诚说，他们更加愿意作古典类的建筑施工，因为古典类的建筑有很多的变化，特别是在转角和墙体与地面和顶棚的交接部位，往往可以用古典的花边线脚达到遮丑的目的。看来如何在施工中达到"横平竖直"的要求，看上去简单，而实际上却还非常遥远。

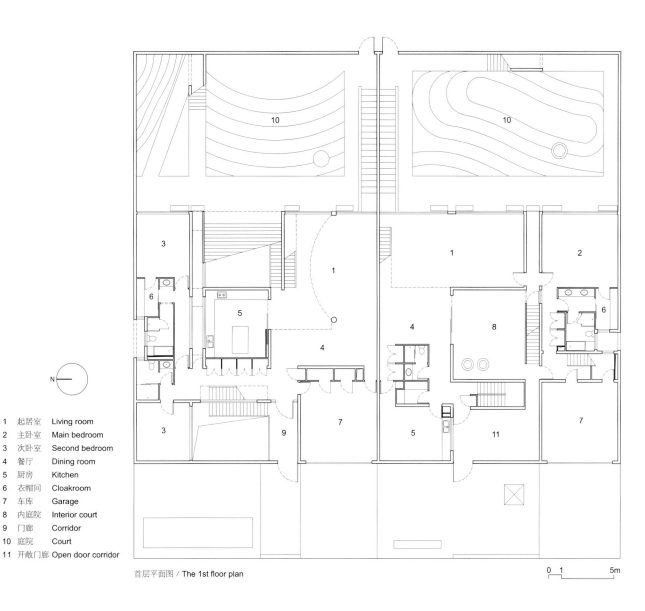

首层平面图 / The 1st floor plan

1 起居室 Living room
2 主卧室 Main bedroom
3 次卧室 Second bedroom
4 餐厅 Dining room
5 厨房 Kitchen
6 衣帽间 Cloakroom
7 车库 Garage
8 内庭院 Interior court
9 门廊 Corridor
10 庭院 Court
11 开敞门廊 Open door corridor

Mixed by seven colors, white is an inclusive color and includes richness. It's a disappearing color, namely "virtual scenery clears" then "empty scenery appears". The specific dimension and distance feeling disappears in the white space. This disappearing feeling generated by the white interface can restore the pure ideas and imaginations of space sceneries. With the overturning and floating non-distance feeling experiences, people can finally get the experiences of "where there is no picture is a wonderful scenery".

House A+B composes is located at area of Lushi Villa around Badachu Park, Xishan, Beijing. The villa area has 52 houses Included House A+B. Because the space inside shows "rigid" and "elastic" characteristics, House A+B is called "prince" and "queen".

House A+B is the largest house in the villa area of which a common one is about 300m²~400m². The total building area is about 700m². Two courts of 18m width and 12m length are extended from the east of the house. The house is composed of two 18m-wide and 7m-high square boxes, two floors aboveground and one floor underground. The abstract house is expected to bring more possibilities to the future residents.

The house tries to construct space by using abstract white boxes. In internal organization, the designer works hard to express imaginative sceneries by abstract objects. Thus the objects could associate people with their own experiences and enable them to get the higher level intuitive consciousness.

It is difficult to complete the two seemingly simple buildings. At first, we faced the construction quality proplems of how to realize "horizontal even vertical" in actual construction. When we communicate with workers, they honestly speak that they are willing to construct the classic buildings because these buildings change much and the classic flower edge lines can cover the flaws. How to realize the requirement of "horizontal even vertical" seems to be simple, but it is very hard to achieve.

二层平面图 / Second floor

1	起居室	Living room
2	主卧室	Main bedroom
3	次卧室	The second bedroom
4	书房	Study room
5	衣帽间	Cloakroom
6	健身房	Gymnasium
7	走廊	Corridor
8	阳台	Balcony
9	室外平台	Outdoor platform

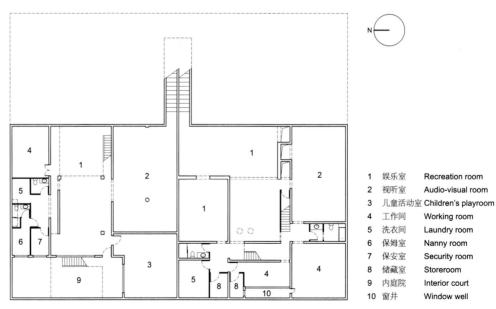

地下一层平面图 / Underground floor

1	娱乐室	Recreation room
2	视听室	Audio-visual room
3	儿童活动室	Children's playroom
4	工作间	Working room
5	洗衣间	Laundry room
6	保姆室	Nanny room
7	保安室	Security room
8	储藏室	Storeroom
9	内庭院	Interior court
10	窗井	Window well

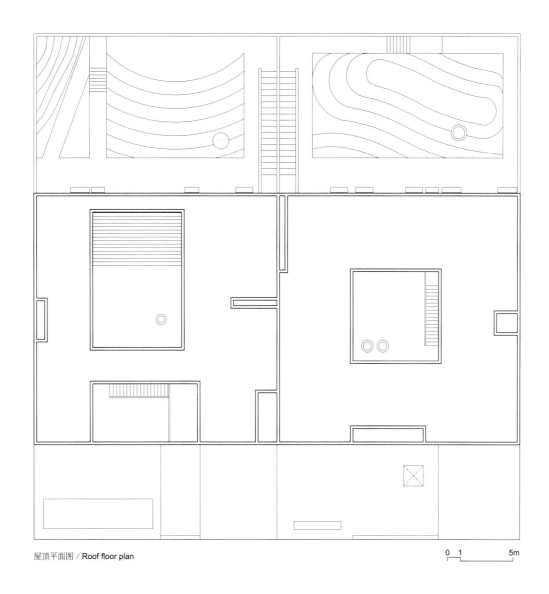

屋顶平面图 / Roof floor plan

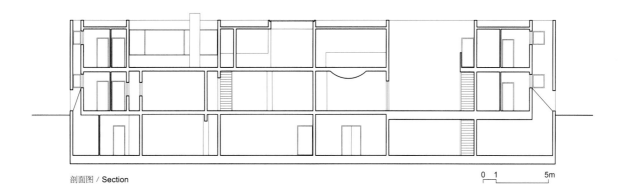

剖面图 / Section

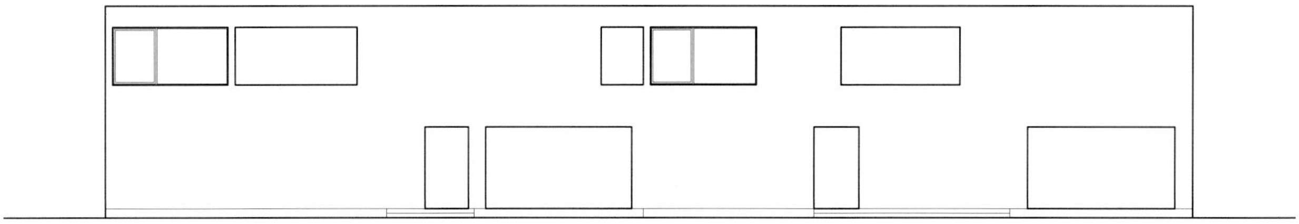

西立面 / West elevation

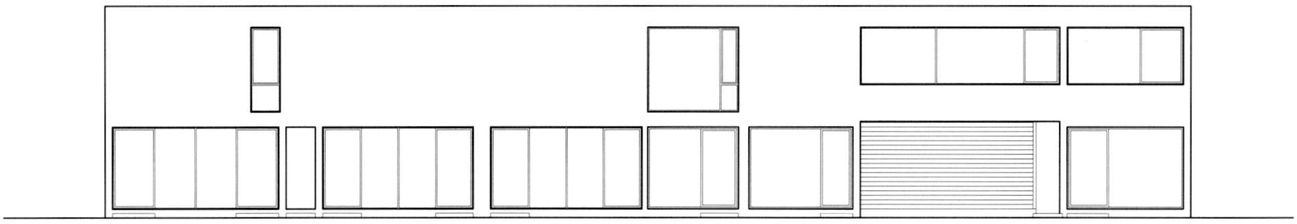

东立面 / East elevation

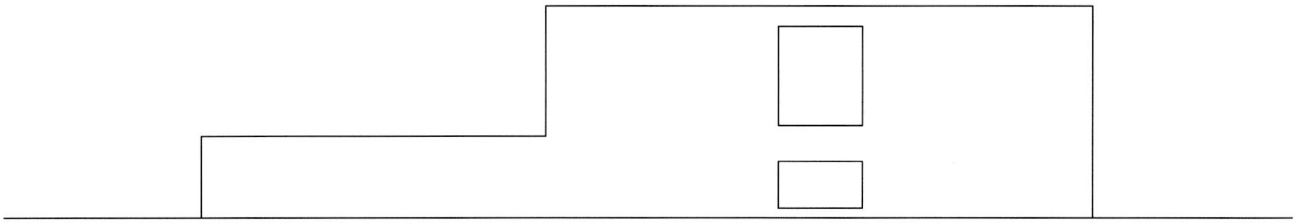

北立面 / North elevation

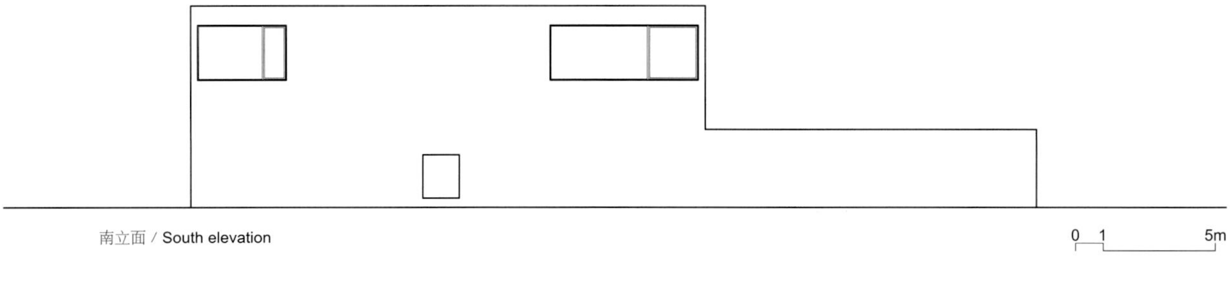

南立面 / South elevation

0 1   5m

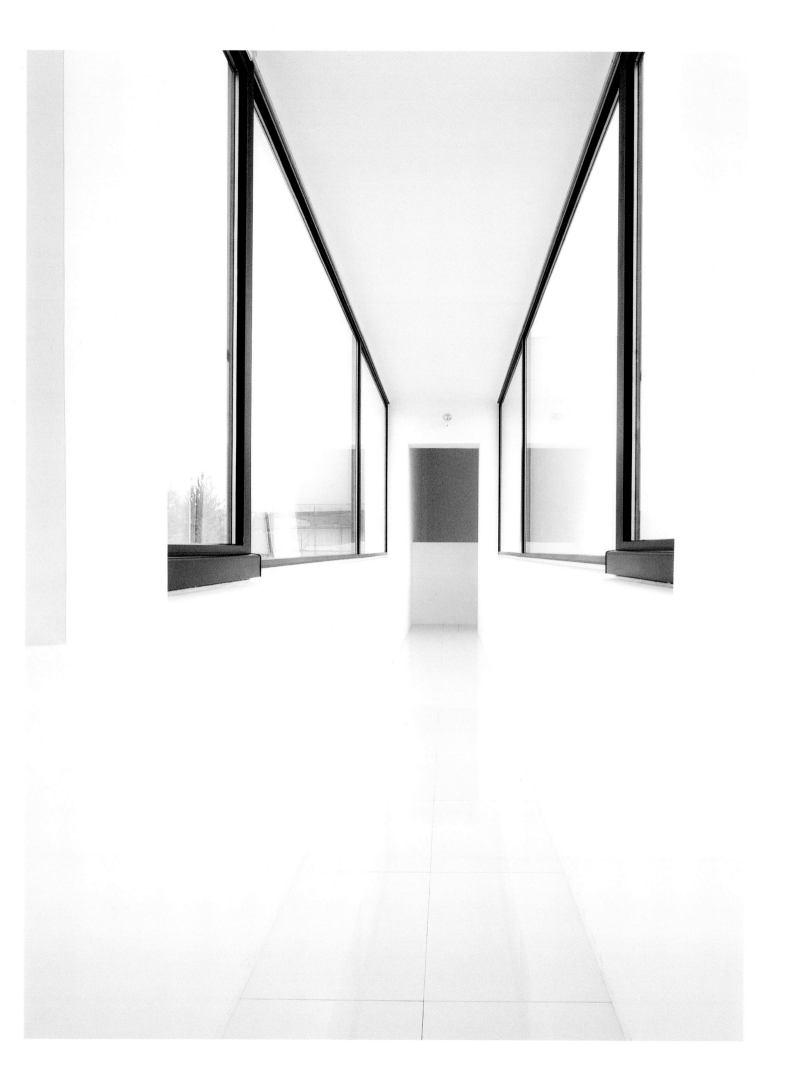

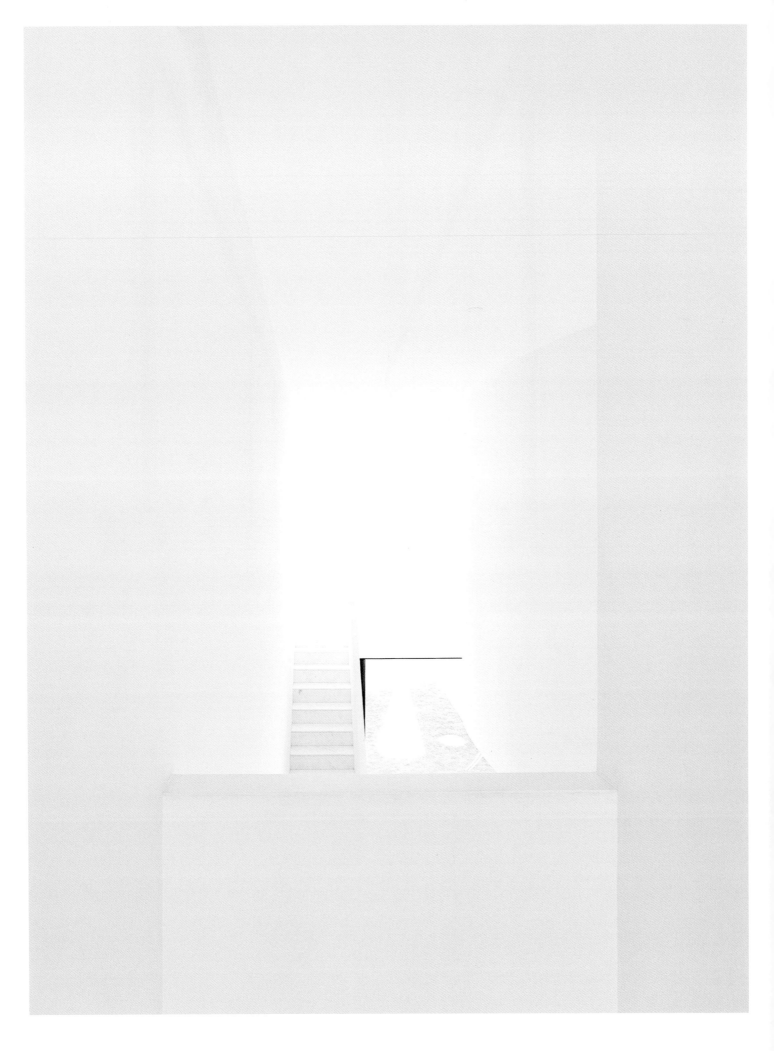

庐师山庄会所 北京
The Lushi Hill Club, Beijing, 2003

这是坐落在北京西郊的一个住宅区会所,为区内的52栋别墅提供一个公共活动的场所,建筑面积为1600平方米,建筑共3层,内部整体拟设置咖啡厅、会议室和4间客房。会所的入口是一个3层高的大厅,大厅与室外以整面的玻璃幕墙相隔,室内与室外之间互为对象的对景与互位关系是该建筑设计中的主旨。

会所西向面对的是整个区域的中心绿地,这个绿地保存了基地上原有的园林树木,具有良好的景观。为此,沿着面向中心绿地景观的展开面方向,设计时在大厅设置一个直接通向二层的坡道,这个坡道能让人们在逐渐上升的过程中体验西侧大玻璃幕墙之外逐渐升起的景象。会所建筑在整个的施工过程中充满难点,入口处3.6米宽、3.6米高的门扇的完成过程,经历了与几年前善美门厅大门设计与制作同样曲折的过程。而西侧整体采用吊挂系统的高11米的幕墙,由于幕墙上部分的玻璃有7米高,并且根据规范需要钢化,现实中进行钢化玻璃的钢化炉只能提供6.8米的玻璃,使得这个幕墙无意中挑战了这个时代的极限。

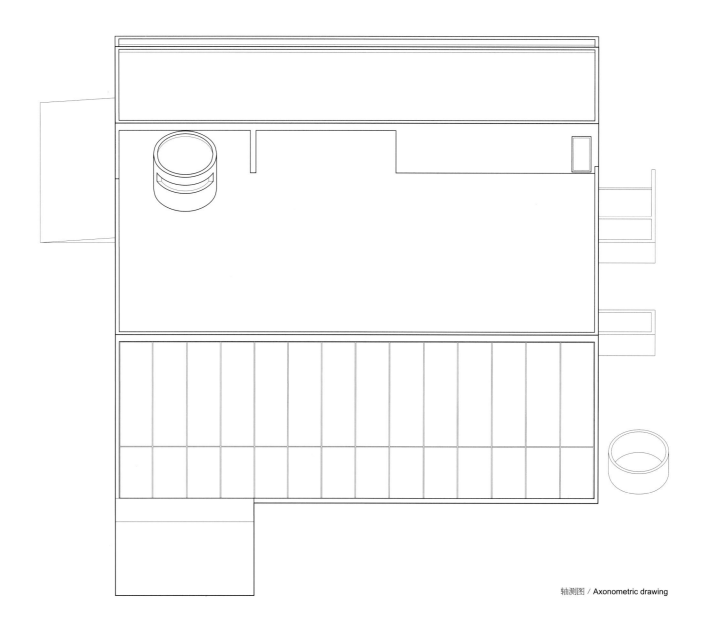

轴测图 / Axonometric drawing

This club is located in a residential area in the west outskirts of Beijing and is a activity site for 52 villas there. The building area is 1600m². This building includes three floors and consists of coffee halls, meeting rooms and four guest rooms. A three-floor hall at the entrance of the club is separated from the out by a whole glass curtain wall. The main point of this building is the vesion relations between the outdoor and indoor objects. The club faces to the center green land of the whole area which reserves old trees with nice sceneries. Thus a slope connected to F2 in the hall is designed to enable people to experience the sceneries outside the glass wall. There are many construction difficulties of this club. The 3.6m-wide and 3.6m-high door at the entrance experienced difficulty in designing and making phase. The 11m-high. glass wall is suspended in the west. The upper glass wall is 7m high and need to be tempered according to the standard, but only 6.8m tempered glass can be produced by the tempering furnace, thus the curtain wall challenges the limit of this era unintenionally.

1	大厅	Hall
2	办公室	Office
3	服务间	Service room
4	茶水间	Tea room
5	保安室	Security room
6	消防控制室	Fire control center
7	煤气表间	Gas meter room
8	垃圾收集间	Rubbish collection room
9	客房	Property room
10	内庭院	Interior court

首层平面图 / The 1st floor plan

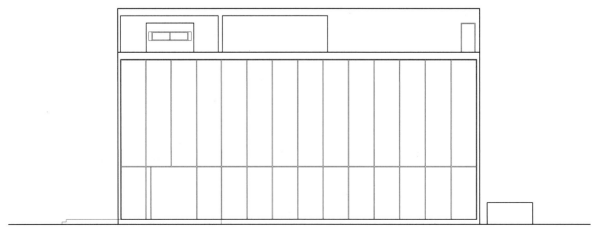

西立面图 / West elevation

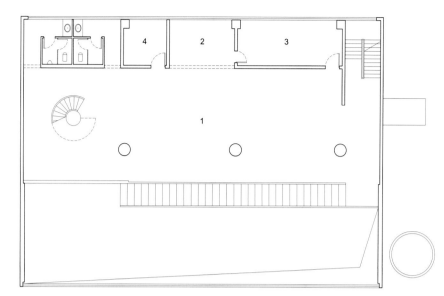

二层平面图 / The 2nd floor plan

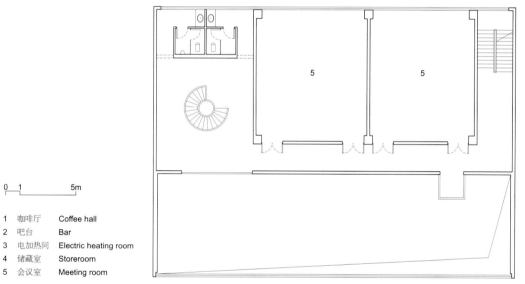

0 1   5m

1 咖啡厅   Coffee hall
2 吧台   Bar
3 电加热间 Electric heating room
4 储藏室   Storeroom
5 会议室   Meeting room

三层平面图 / The 3rd floor plan

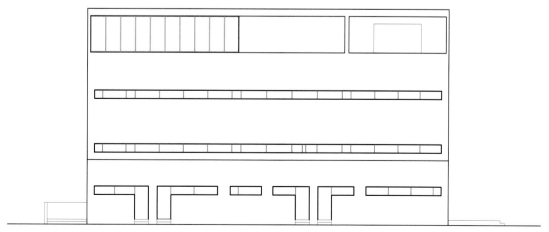

东立面图 / East elevation

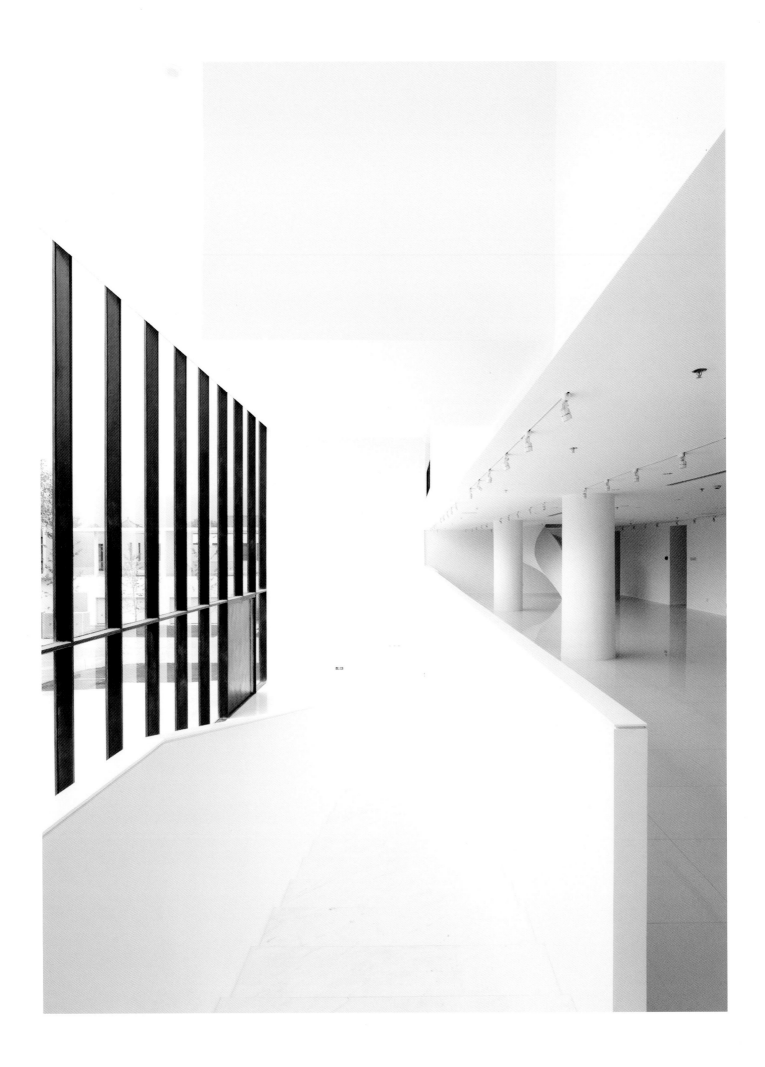

百子湾小区中学 北京
BaiziBay District Middle School, Beijing, 2003

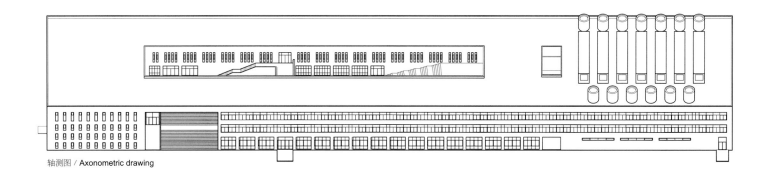

轴测图 / Axonometric drawing

百子湾中学校位于北京的东四环外，是一个为60万平方米住宅社区服务的24班中学校舍。学校占地2.57万平方米，基地沿东西方向展开。在整体的设计布局时，将教学楼沿东西方向布置，同时在其南侧布置一个400米的标准操场。"一"字形的校舍全长约157米，建筑总面积为1.1万平方米，建筑整体设计为3层。学校一层布置有实验室、图书室、行政人员办公室以及教师和学生食堂。二层部分安排了一个由教室围合的长条形内院式中庭，坐落在一层屋顶上，中庭内设置有给一层提供采光的采光筒。同时这个二层的中庭还设置有通往三层的大台阶和楼梯，从而使得整个中庭成为学校内部的内院广场。在课间，学生们自由地从各个方向汇聚在这里，大家在这个院子里进行交流、玩耍和休息，同时也可以进行集会活动。这个空间和学校外面保持相对独立性，学生在里面不会受到学校外面的干扰。而这样的中庭内院式的空间意向，从高层的住宅上空俯瞰，实际上也在唤起人们对于中国西北地区下沉式窑洞空间的联想。学校的主入口在北侧，入口上空有一个跨度为16米的空中连桥，目的是使三层的交通形成一个环路。空中连桥的下方是一个大台阶，直通二层的中庭内院，从而使二层的中庭内院又如一个开敞门厅起到分流和聚合的作用。学校西侧二层设有室内运动场，运动场的下部是学生与教师的食堂与餐厅。值得一提的是：运动场上空的一组天窗在给室内运动场采光的途中扫过不同的区域，在给运动场采光的同时也给运动场与画室之间的走廊带来了光线。

The Baiziwan middle school is outside the east 4th Ring Road of Beijing and includes 24 classes which serves for about a residential quarter 600000m². The base occupies 25700m² and extends from west to east. We deploy the main building along the base and a 400m standard playground in the south. The "——" type building is about 157m long. The total building area is 11000m². The building is three floors. The labs, laboratories, administration offices and dining rooms are at F1. The quadrate internal court on the top of F1 is enclosed by classrooms. Light canisters for F1 are in the cortile which connect to F3 by stairs and big steps, thus the cortile becomes the inner court square of the school. Students can converge here freely from different directions and communicate, play and relax here. This courtyard space can keep students free from disturbances of outside. The inner court space may inspire peoples' associations of the sunken cave space in North-West area of China when looking down from the high buildings around. The entrance of the school is on the north, a 16m overhead bridge is above to form a circuit traffic of F3. A big steps under the bridge connects to the cortile at F2, so the cortile at F2 likes an open door hall to distribute and associate traffic flow. A indoor playground is at F2 in the westand dining room is under it. It's noticeable that a group of skylights above the playground can scan different areas when they light the indoor playground, and bring light for the corridor between the playground and art studio.

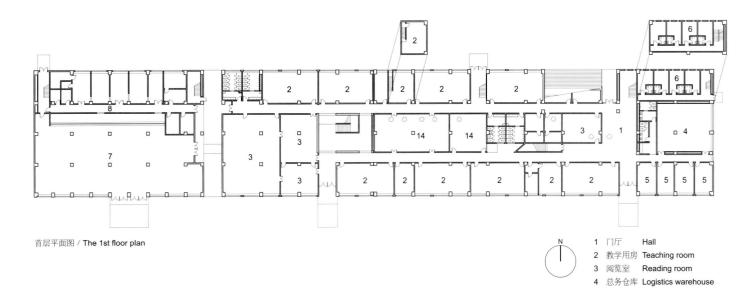

首层平面图 / The 1st floor plan

1 门厅	Hall
2 教学用房	Teaching room
3 阅览室	Reading room
4 总务仓库	Logistics warehouse

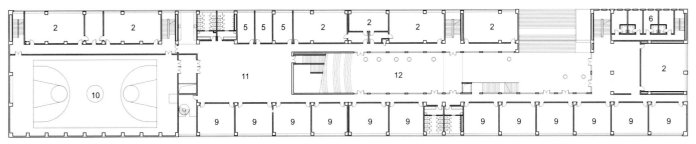

二层平面图 / The 2nd floor plan

5 办公室	Office
6 教室宿舍	Classroom and dormitories
7 餐厅	Dining room
8 厨房	Kitchen

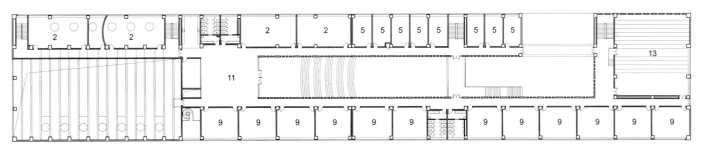

三层平面图 / The 3rd floor plan

9 普通教室	Common classroom
10 风雨操场	Open playground
11 室内休息区	Indoor relaxation area
12 室外活动平台	Outdoor activity platform
13 阶梯教室	Terrace classroom
14 展览空间	Exhibition space

南立面 / South elevation

0 1 5m

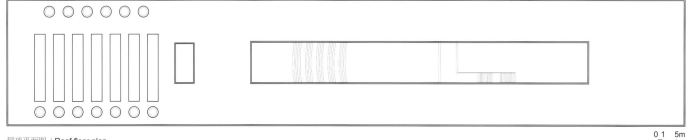

屋顶平面图 / Roof floor plan

北立面 / North elevation

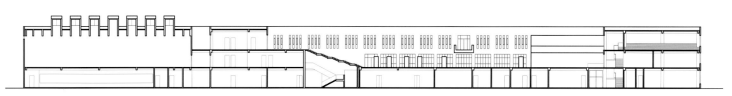

南侧剖面图 / South section

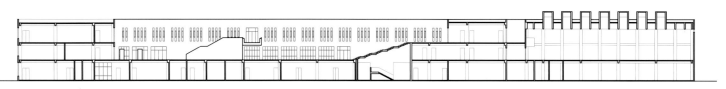

北侧剖面图 / North section

西立面 / West elevation

东立面 / East elevation

百子湾小区幼儿园 北京
BaiziBay District Kindergarden, Beijing, 2003

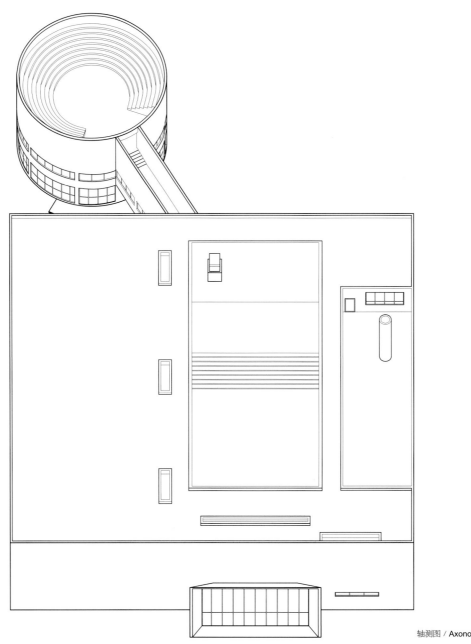

轴测图 / Axonometric drawing

    北京百子湾幼儿园与百子湾中学校一样，也是为百子湾小区居民服务的公共设施，幼儿园占地面积4000平方米，总班级数为9个班。该幼儿园的总建筑面积为3200平方米，为3层。整个建筑是由一长36.2米、宽37.5米、高为10.5米的方形体块与一个半径为7米、高为10.5米的圆柱体块组合而成。作为建筑主体的方形体块共3层，入口处的"喇叭"让孩子们有"小喇叭开始广播了"的联想。进入"喇叭"，是一个三层高的入口大厅，大厅的上空设有一个窄长条的天窗将阳光导入，伴随时间的不同，条形的光线能够在中庭的墙面上形成不同的光与影的交融。入口大厅的左侧是教室区域，右侧是厨房和办公区。在幼儿园的南侧区域集中布置了活动室和孩子们的卧室。北侧则在一层布置厨房和办公室，二层布置一个图书室。三层布置有一个露天的活动场地。从作为幼儿园主体的方块体量上看，结合南北两侧区域之间的分区，体块的中心设有一个内院，从而使得方形体块形成一个"虚中"的意向，并迎合着"不虚不足以妙万物"的中国古代哲学思想。而这个"妙万物"之地，在实际使用中是一个露天的大厅。这里白天可作为孩子们的室外露天活动场地，晚上可以作为室外露天的放映厅。3层高的圆柱体位于建筑的西南方向，本身是作为一个独立的体块而存在。其二层是一个圆形的音乐教室，教室的屋顶是一个露天小剧场。从主体的方形建筑体块的二层与三层分别可以通过一个狭窄的室内和室外的走廊通达到这个二层的音乐教室和三层的露天小剧场。幼儿园建筑采用色谱完整的白色，以突出建筑整体的"积木"感。之所以不在这里采用大红大绿的色彩涂抹，是为了避免由于片面地采用某种强烈的颜色在儿童成长过程中对其造成片面的精神刺激与视觉伤害。

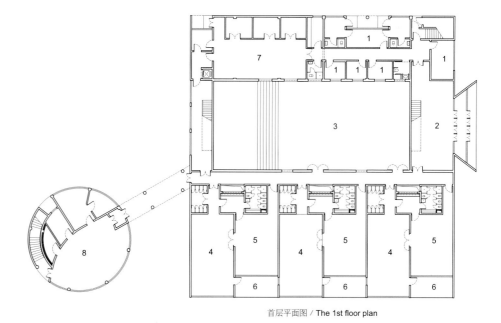

首层平面图 / The 1st floor plan

1	办公室	Office
2	门厅	Hall
3	内庭院	Interior court
4	活动室	Playroom
5	寝室	Bedroom
6	阳台	Balcony
7	厨房	Kitchen
8	婴儿室	Feeding room
9	阅览室	Reading room
10	音体教室	Music classroom

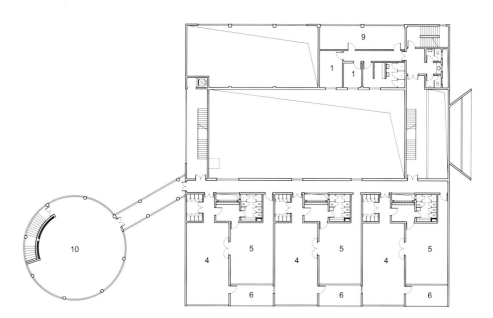

二层平面图 / The 2nd floor plan

Like Baiziwan middle-school, the Beijing Baiziwan kindergarten is also a public facility that serves for the residents in the Baiziwan residential quarter. It occupies 4000m² and includes 9 classes. The total building area is 3200m². The whole building is a 36.2m²-long, 37.5m²-wide and 10.5m²-high quadrate box and is combined with the 10.5m²-high column of 7m radius. The quadrate box as the main building includes three floors. The "horn" at the entrance gives children the impression of "it starts to broadcast".

Enter the horn, it's a F3 entry hall with a narrow skylight above it. With time changing, it shows the performances of lights and shadows on the wall of the cortile. The classroom is on the left of the entrance hall and the kitchen and office area is on the right. The activity rooms and bedrooms are in the south. In the north, there are kitchen and office area at F1, a laboratory at F2 and an open activity site at F3. A courtyard between the south and north area forms a "void center" which echoes the traditional Chinese idea of "if there is no empty, everything can not be wonderful". The "empty" here is an open hall which acts as activity site on the day and open project hall on the night. Three-floor column is in the southwest as an independent part. A circle music classroom is at F2 and an open small theater is on the top. A corridor connect this part with the main building. The buildings are painted white to creat the "building block" feeling. Strong colors are not uesed to prevent partial spirit spurs and visual injuries to children.

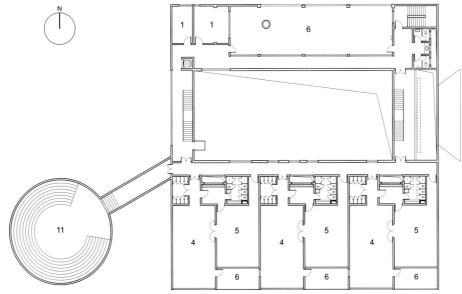

1	办公室	Office
4	活动室	Playroom
5	寝室	Bedroom
6	阳台	Balcony
11	露天剧场	Open theater

三层平面图 / The 3rd floor plan

西立面 / West elevation

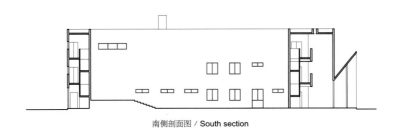

南侧剖面图 / South section

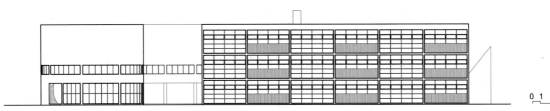

南立面 / South elevation

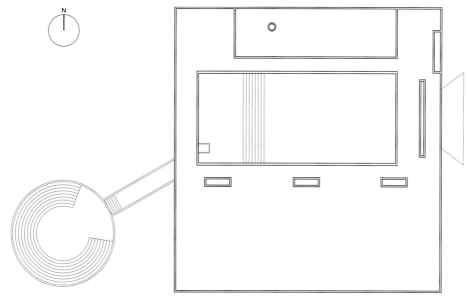

屋顶平面图 / Roof floor plan

东立面 / East elevation

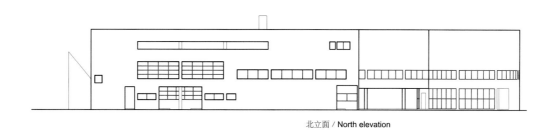

北立面 / North elevation

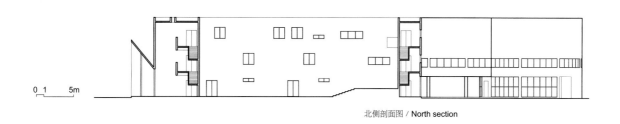

北侧剖面图 / North section

石景山财政培训中心 北京
Shijingshan Financial Training Centre, Beijing, 2003

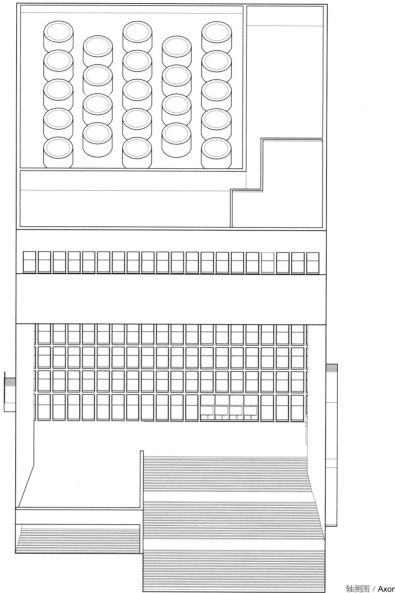

轴测图 / Axonometric drawing

　　北京石景山财政局建筑位于北京的石景山区，西邻石景山区法院建筑，东侧为一个商业用地。项目的基地是一个南北方向的条形用地，总建筑面积为7000平方米。在使用功能的安排上，设计时将有对外功能的财政营业大厅放在南侧，而内部人员的出入口设置在北部，从而决定了整个建筑的主体流线关系。考虑到未来办公人员的舒适度，设计时没有采用"一条走廊两侧排房间"的布局，而是将所有的办公室门与走廊面向一个巨大的中庭，目的是使得每一个工作人员在办公室的出入之间获得"收"与"放"的心情调节和气氛转换。基于这样的理解，建筑被设计成宽与高均为40米，长为60米的体块，其内部被嵌入一个长和宽均为20米，高为38米的室内中庭。位于中庭上部及其西侧墙面上的46个3米直径的圆形采光孔，为中庭内部倾泻阳光的同时还突出了"凿"的概念，而这种"凿"的概念本身还是中国传统哲学思想的体现。中国古代哲人老子所谓"凿户牖以为室，当其无有室之用"的论述，使该建筑本身在"凿"的概念上获得地域性意义的解读。

This Building is located in Beijing Shijingshan District and is adjacent to the Shijingshan Court in the west and a business land in the east. Based on a quadrate land, the total building area is 7000m². The open business office hall is placed in the south and the exit/entrance of the staffs in the north, thus the main flow line is determined. Considering the comfort of the staffs in future, the layout of "one corridor with rooms on both sides" is abdicated. All officeroom doors and the corridor face to a huge cortile to offer a comfortable ambience. Based on this point, the building is designed as a 40m×40m×60m cube with a 20m×20m×38m cortile embedded. 46 round light holes of 3m diameter are above the cortile and on the west wall to highlight the concept of "chisel", as Lao Tze states "The door and windows are chisel out (from the walls) to form an apartment; but it is on the empty space (within), that its use depends".

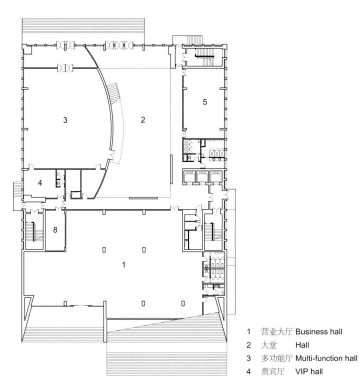

1 营业大厅 Business hall
2 大堂　　Hall
3 多功能厅 Multi-function hall
4 贵宾厅　VIP hall

首层平面图 / The 1st floor plan

二层平面图 / The 2nd floor plan

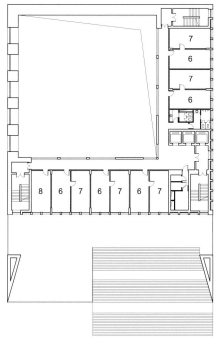

5 培训教室 Training classroom
6 科员室　Staff room
7 科长室　Section director room
8 休息室　Rest room
9 休息大厅 Rest hall

标准层平面图 / Standard floor plan

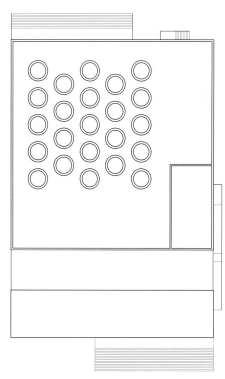

屋顶平面图 / Roof floor plan

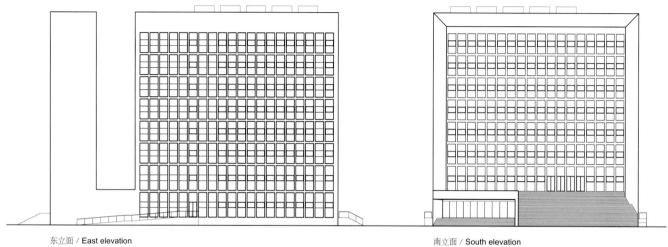

东立面 / East elevation　　　　南立面 / South elevation

0 1　5m

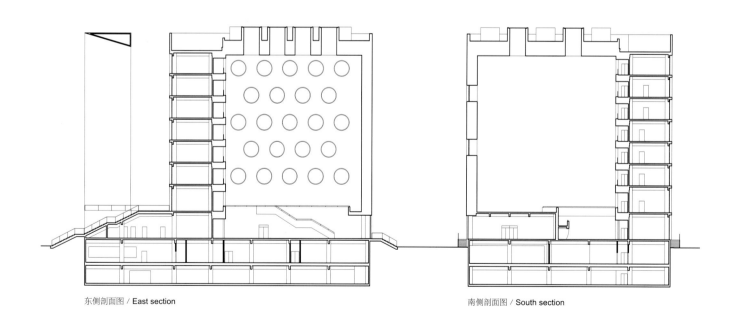

东侧剖面图 / East section　　　　南侧剖面图 / South section

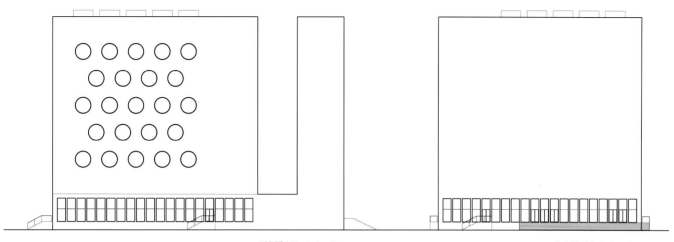

西立面 / West elevation　　北立面 / North elevation

0 1　5m

西侧剖面图 / West section　　北侧剖面图 / North section

茵莱玻璃钢门窗制品有限公司办公楼 北京
Inline FRP Doors and Windows Products Co., Ltd. Office Building, Beijing, 2006

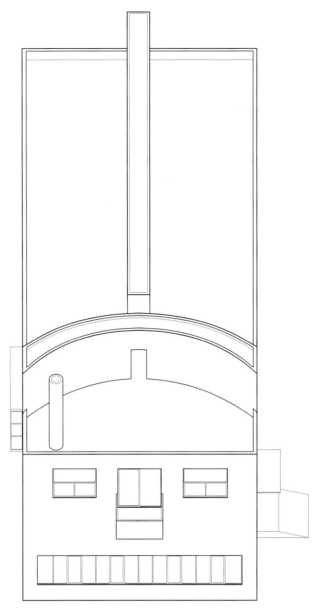

轴测图 / Axonometric drawing

这是为茵莱玻璃厂设计的办公楼项目,位于北京市通州区,总建筑面积为1579平方米。方案在设计初始被要求要尽可能地使用玻璃窗,因为作为玻璃窗厂的办公楼,使用玻璃窗本身就是对于产品的一种展示,与此同时,更多地采用自家生产的玻璃窗,在整个造价预算上也会有所降低。本着这样的目标,也根据甲方要求的低造价的原则,建筑的整体造价最终控制在200万元人民币左右。

由于场地的主要入口位于东侧,因此整个建筑采用南北方向的布局。主要入口位于建筑的东侧,南部面向并正对玻璃窗生产车间的出入口,而车间的出入口恰好能够框出该办公建筑的正南面,出入之间能够使整个建筑令人感受到"虎虎生威"的建筑意向。

建筑整体采用一个长38米、宽15米和高为14.2米的长方形箱体。箱体建筑整体为3层,入口大厅是一个三层高的吹拔空间。在这个空间中,充满了交错的通往各个部分的楼梯和通道,映射着皮拉内西绘画中的复杂结构。这个箱体建筑,还由入口大厅形成的三层高吹拔空间将整个办公楼分为南北两个区域。箱体南侧的一层是一个接待用的会议室,二层是一个大型会议室,而三层则是总经理的办公室。

箱体的北侧一层是一个产品的展示厅,在展示厅的后部,布置有办公室和两个小的客房供加班的员工临时居住。二层是主要的办公区,从这里员工们可以通过办公区一侧的楼梯直接通达到一层的产品展示室和三层的管理层办公室。从二层的办公区到三层管理层办公区之间设有一个向上的坡道将二者相连。同时在这个坡道的上空,还设计有一个条形的玻璃天窗。实际上由于这是一个玻璃窗生产厂的办公楼,因此在进行设计的过程中如何展示玻璃窗在应用中的可能性是一个重要的主题,同时通过玻璃窗的应用,展示光的魅力也尤为重要,而整个建筑整体的物理透明性表达,以及建筑上部的条形天窗的表现恰是对于这样一个主题的回应。建造过程中,施工质量的问题一直是对这项工程的一个困扰,能否使建筑产品做到"横平竖直",仍然是当下所面临的一个难题。

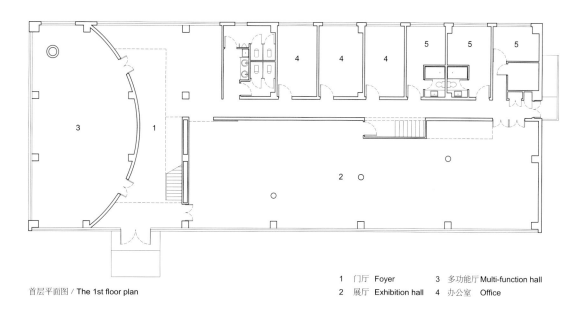

首层平面图 / The 1st floor plan

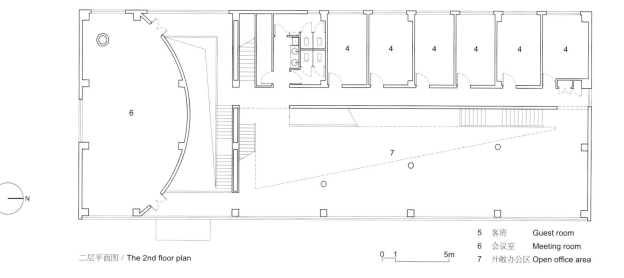

二层平面图 / The 2nd floor plan

1	门厅	Foyer	3	多功能厅	Multi-function hall
2	展厅	Exhibition hall	4	办公室	Office
5	客房	Guest room
6	会议室	Meeting room
7	开敞办公区	Open office area

This building is designed for Inline glass factory and is located in Tongzhou District of Beijing. Total building area is 1579m^2. To use as much as possible glasses products of the factory itself is asked by the owners in the initial design phase to exhibite their products and reduce cost budget. For this target, the whole cost of the building is controlled within 2 million Yuan according to low cost rule of the owners.

We deploy the building from south to north and the main entrance in the east. Follow the main entrance of the site, the south faces to the exit / entrance of the glass window plant. The exit / entrance of the plant can frame the south face of the office building.

The whole building is designed as a 38m-long, 15m-wide and 14.2m-high box with three floors. The entrance hall is a F3 patio space where crossed stairs and passages connect to different parts. Its complicated structure is like the Piranesi image. The entrance hall also divides the whole office building into the south and north part. The reception meeting room is located at the south F1. A large meeting room is at F2. The general manager office room is at F3.

The products exhibition hall is at the north F1. Behind it are offices and two guest rooms for the overtime staff to temporarily live. The main office area is in the north F2 where staffs can pass the stair beside to directly reach the exhibition room in F1 and the management offices in F3. An upward slope connects the F2 office area to F3 management area with a skylight on the ceiling. In fact, how to exhibit the possible applications of glass windows and the charm of light is significant. The physical transparency of the whole building and the skylight above both reflect this subject. Construction quality is troublesome in engineering, so it is still difficult to guarantee "horizontal even vertical" in the building.

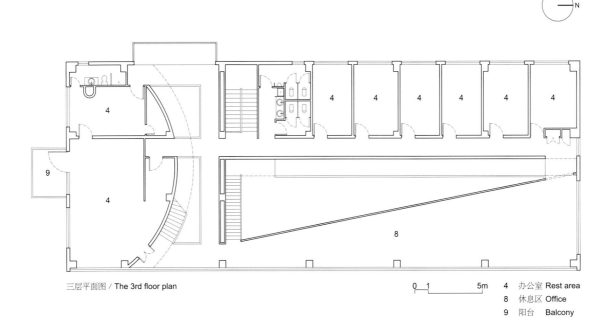

三层平面图 / The 3rd floor plan
0 1   5m
4 办公室 Rest area
8 休息区 Office
9 阳台   Balcony

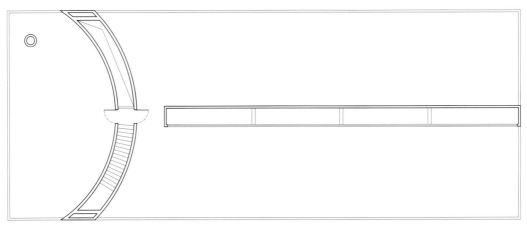

屋顶平面图 / Roof floor plan

南立面 / South elevation   北立面 / North elevation

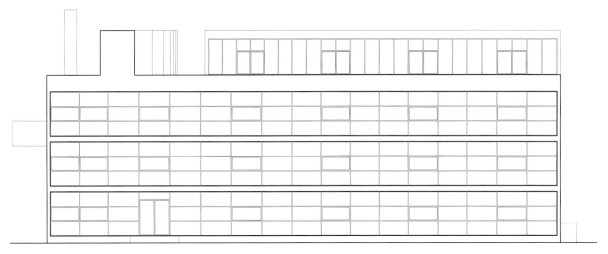
东立面 / East elevation

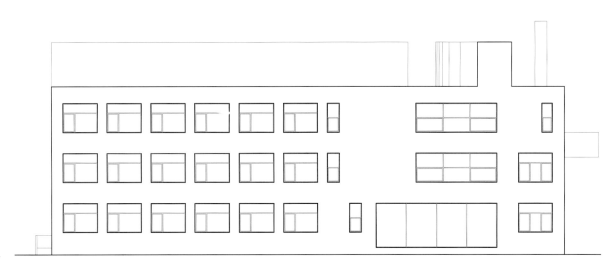
西立面 / West elevation

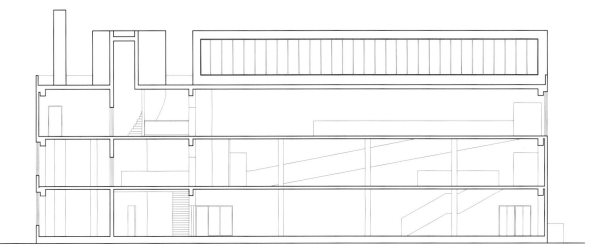
东侧剖面图 / East section

国轩国际研发中心商业设施 北京

Commercial Facilities of Guoxuan International R&D Centre, Beijing, 2005

这是一个商场设计项目，位于北京市通往机场的高速路北皋出口的北侧。考虑到区域周边已经建成大量的别墅，缺少有规模的商业配套设施，于是业主决定在这个区域内建设一个商业建筑。但是作为商业建筑的地点，这个地段有一个特殊性，即在商业道路的西侧有一个医院，如何保证商场空气的流通便成为思考这个商业建筑设计的根本出发点。结合该地区对于小型商铺的业态需求，不采用一般意义上的综合商业楼，而是让整个建筑由不同的小商铺构成的一个商业聚落综合体的构思便成为该设计的主要特征。这个商业建筑共4层，总建筑面积为14287平方米。建筑整体长度为146米，宽度为33米。建筑的一至三层是由外廊围合而成的开敞空间，四层是一个漂浮在柱廊上部封闭的箱体，并作为餐饮空间使用。考虑到商铺的独立运作的灵活性，所有店铺均独立地面向公共交通走廊。而所有的交通连接要素均展现在最外侧柱廊的开敞部分，楼梯、坡道、自动扶梯等这些输送装置源源不断地直接将人们输送到他们想要到达的位置。

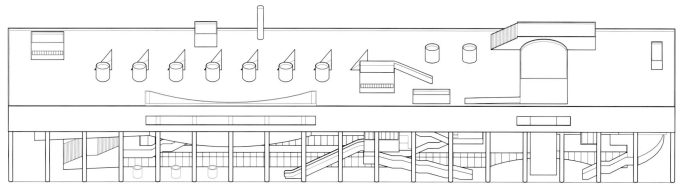

轴测图 / Axonometric drawing

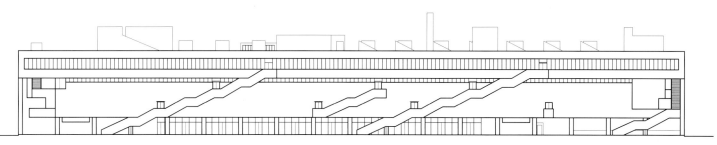

北立面 / North elevation

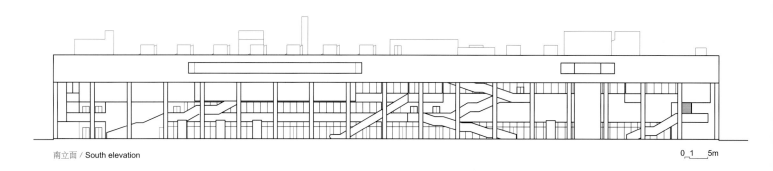

南立面 / South elevation

0 1 5m

This market project locates in the north of Beigao exit of the express way to the airport in Beijing. Considering many villa areas there without available large commercial facilities, the owners decide to construct a business building in this area. A hospital in the west of the commercial road make the site special. Thus how to guarantee the air ventilation of the marketplace become crucial. Out of the demand of small commercial stores of this area, a complex composed of different small commercial stores become the main feature instead of the common comprehensive commercial buildings. This project includes four floors. Total building area is 14287m². This building is 146.7m long and 33.3m wide. The F1-F3 are open spaces enclosed by outside corridor. F4 is an enclosed box floating above the corridor for catering. Considering independent operations and flexibilities, all stores are independent and face to the public traffic corridor. All transport devices displayed in the corridor such as stairs, slopes and escalators continuously transport people to their destinations.

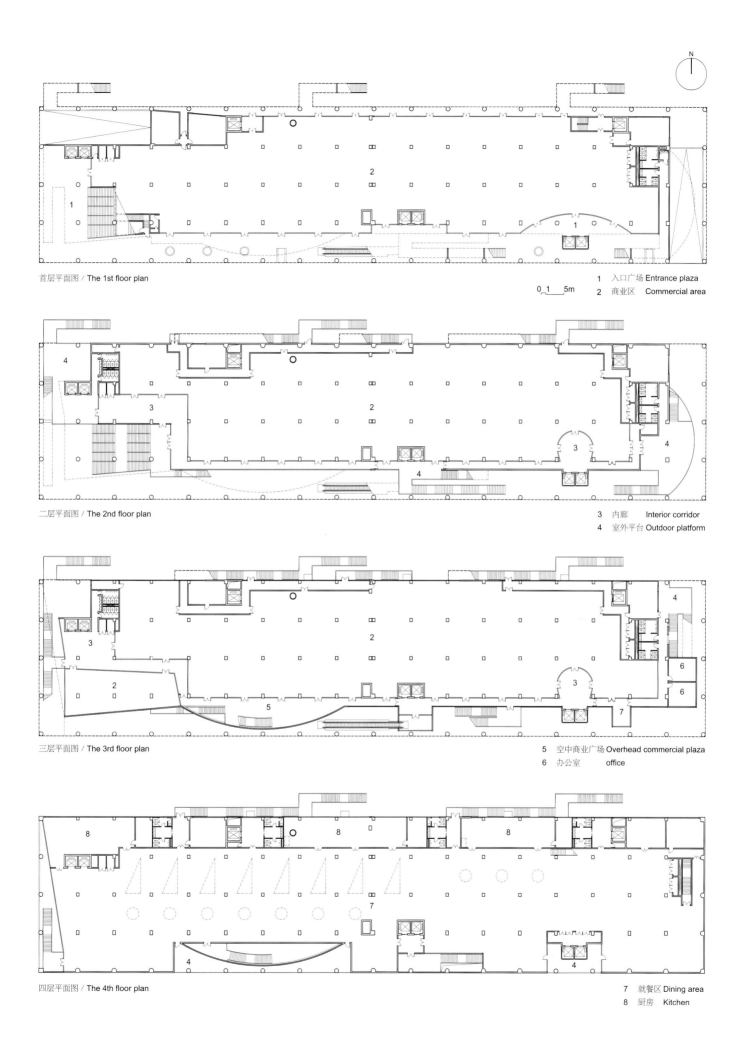

首层平面图 / The 1st floor plan

1 入口广场 Entrance plaza
2 商业区 Commercial area

二层平面图 / The 2nd floor plan

3 内廊 Interior corridor
4 室外平台 Outdoor platform

三层平面图 / The 3rd floor plan

5 空中商业广场 Overhead commercial plaza
6 办公室 office

四层平面图 / The 4th floor plan

7 就餐区 Dining area
8 厨房 Kitchen

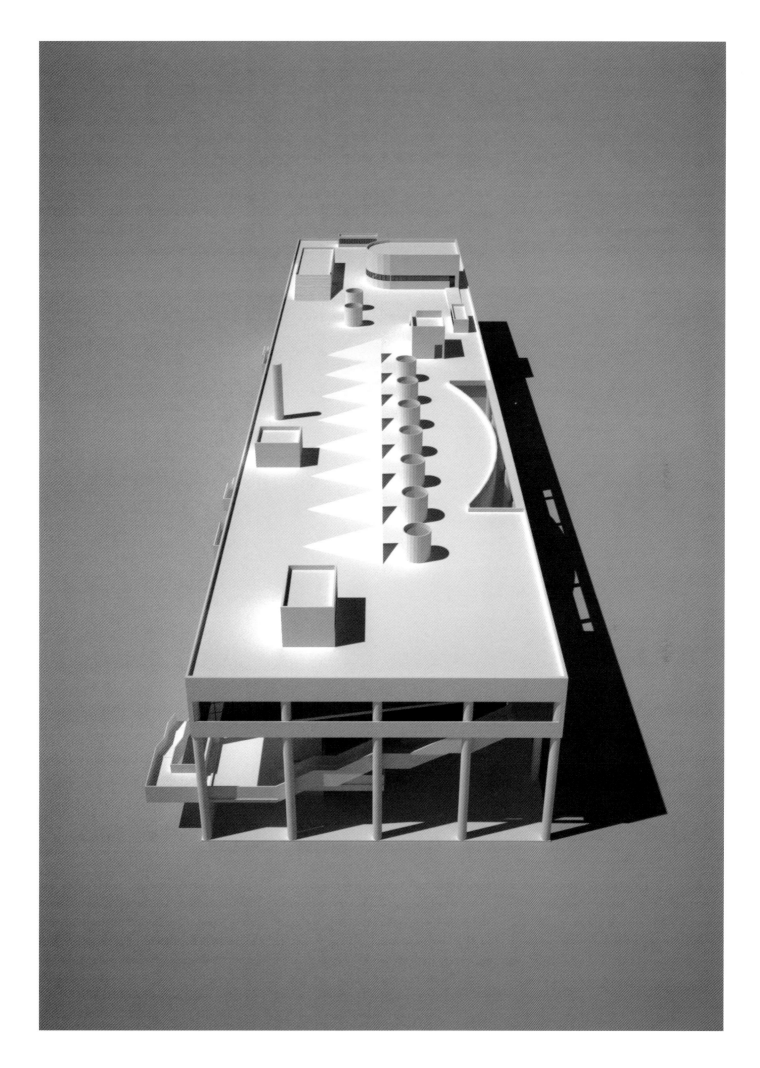

柔软住宅 北京
Soft House, Beijing, 2008

住宅是自由的，可以根据不同的条件和场景来进行布局。这个柔软住宅实际上是对于住宅的柔软性理解的一种探讨。有的人希望将自宅装点得堂皇而显示荣耀，有的则隐逸不显张扬。中国传统中的隐士追求的内敛和内在的丰富取向，实际上与现代建筑理念中注重空间的丰富性的表述是异曲同工的。这个约1000平方米的住宅建筑，坐落在一个平整的草地上，设计时采用一个柔软的墙体将所有的建筑加以围合，墙体本身也是建筑的一部分。也正因如此，墙体变化给予建筑空间的可能性便凸显出来。建筑内部的布局是简单和明快的，功能性的紧凑能够增加使用者的快适程度。住宅中间的走廊上空有一个天窗，天窗采用板材翘起的方式用以控制光线的变化。东侧的靠近入口的位置，伴随天窗采光口的加大而明亮。越往走廊的深处，伴随天窗起翘板的降低，通光量的减少，令人感到空间的私密性也在加强。而这种依靠光亮的调节来进行区域划分的尝试是这个设计的又一特征。住宅入口的处理同样希望采用这样的用光方式取得调解和对比，入口处的筒状构造，力图求得欲扬先抑的空间节奏感。建筑周边以落地的玻璃相围合，外部的曲面墙体会造成一种光与影的微妙渐变。

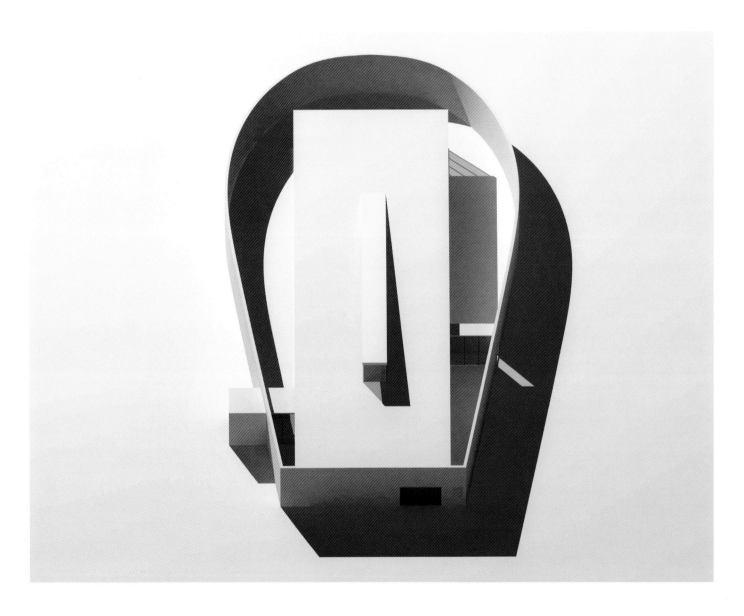

The soft house is free and can be deployed by different conditions and scenes. Some people prefer connotative style when some hope to show glories by their house. Actually, the liability of traditional Chinese anchorets' pursue for reserved style and internal richness is similar with modern architectural concept with emphasis on abundant space expressions. This nearly 1000m^2 house is located on a plain grassland. A soft wall as a part of the building encloses the house and provides more space possibilities. The layout inside is simple and vivid. Functional compaction can enhance comfortabilities. A skylight is above the corridor. The east area around the entrance is brighter, but with descending tilting plate of the skylight, the light is less to enhance the privacy space experiences. Dividing the area by adjusting light is another feature of this house. The entrance of canister structure is another try of this mode to realize the space rhythm feeling of first suppression for later raising. Surrounded by the glass French windows, the outside bending wall will lead to slow changes of light and shadow.

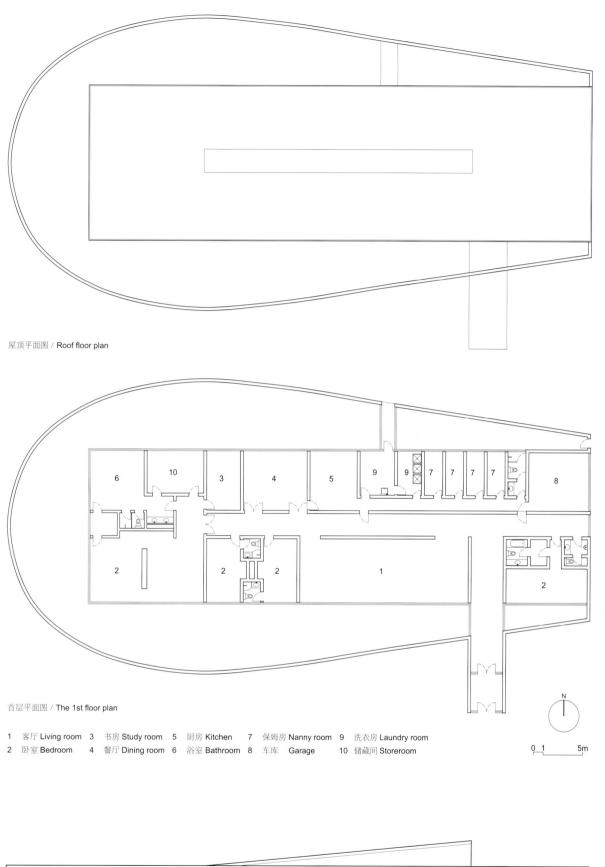

屋顶平面图 / Roof floor plan

首层平面图 / The 1st floor plan

1 客厅 Living room 3 书房 Study room 5 厨房 Kitchen 7 保姆房 Nanny room 9 洗衣房 Laundry room
2 卧室 Bedroom 4 餐厅 Dining room 6 浴室 Bathroom 8 车库 Garage 10 储藏间 Storeroom

南立面 / South elevation

北京万象新天150平方米住宅改造 北京

Wanxiangxintian 150 Square Meters Residential Reformation, Beijing, 2007

这是一个为万象新天住宅小区所设计的住宅室内,该室内面积为150平方米。在设计的过程中试图通过对空间的组织,让使用者获得丰富多变的空间感觉的同时,在有限的空间中给人以超出物理空间之上的扩大空间的感受,是设计的主要意图。一般地,在室内设计的过程中,变动格局是常有的事情。本案的格局变动尝试着能更多地体现使用功能与空间感两方面的互动。在设计的过程中,对于原有的户型平面进行了较大的调整,同时以人的行为出发对功能区进行了整合,去除了部分墙体,整合出具有功能性及趣味性的空间内涵,实现了对于空间的最大利用。尽管新整合的空间没有增加功能区域也没有增加面积,但是通过可变装置和能够延伸视觉的要素细节,将整个空间灵活地融合在一起,从而在使功能灵活可变的同时也使空间变得宽敞。此外,整个室内尝试性地加入了一个弧形的元素,即将原有方案的入口处与客厅相邻的墙面改造成弧形的墙面,进而使入口处产生良好的中途富有变化的视觉进深感。在围绕着中心交通空间的区域,安置了书架格子,增加了主人趣味的展示区域,为爱好书籍和喜爱收藏的主人提供了一个画廊般的展示空间和区域。

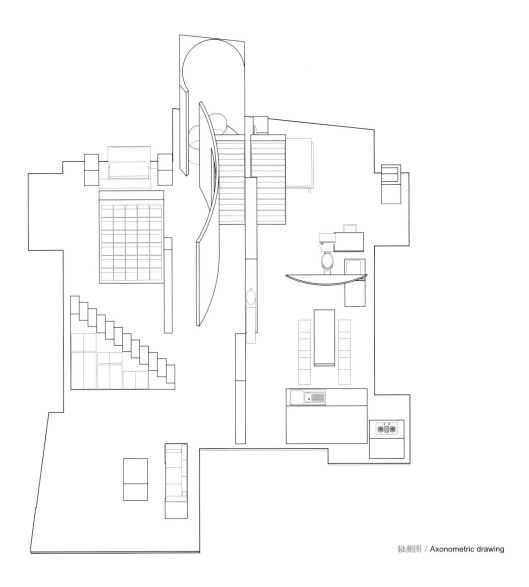

轴测图 / Axonometric drawing

This interier design is for an apartment in Wangxiangxintian residential quarter. The indoor area is 150m². Besides mutiple space experiences, the design aims to provide people an expanded feeling beyond the limited physical area by space organizations. Generally, pattern changing is common in interier design. This design tries more to embody interactions of fuctions and space feelings. The old house plane is adjusted a lot in this design. The function areas are integrated out of human behaviors. Part of walls are removed to form a space of functions and interests and to utilize the space to the most extent. Although the new space does not add functions and areas, it can flexibly fuse the whole space by variable devices and vision-extending elemental details, and thus creat flexiable functions and wider space. Trially reconstruct the wall adjoin the entrance and the guest hall as an arc wall to generate well experiences of vision depth at the entrance. Latticed bookshelves are installed in the center traffic space to provide a exhibition space for the owner who likes books and collections.

苏家坨住宅区小学 北京
Sujiatuo Residential District Primary School, Beijing, 2009

苏家坨小学是一个位于北京北部郊区的社区学校,学校建筑的总体面积为4969平方米。由于学校的用地非常紧张,于是如何处理和整合活动场地的问题就成为设计时首先要思考的问题。此设计中,建筑的主体部分位于基地的南侧,北侧为操场,由于基地的条件限制,操场设为斜向布置。建筑主体长74米,宽24米。主要教室空间位于建筑的南侧,以获得充足的采光。首层建筑面积为1810平方米。北侧首层屋顶为一个室外活动平台,有楼梯与各层空间相连。学生在课间可以轻松到达这里,整个屋顶平台成为开展丰富多彩活动的场所。在这个二层屋顶平台上有直接通向三层体育教室的楼梯以及通达操场的楼梯,从二层的公共教室以及三层的走廊可以直接通达到这里,因此二层的屋顶平台本身也成为了一个公共的小广场。体育活动室、美术教室的屋顶采光带来了更优质的空间感受。整个建筑的流线简洁明确,办公、教学、音乐教室和室内运动场,各部分在不互相干扰的前提下通过不同空间要素的转换被紧密地联系在一起。

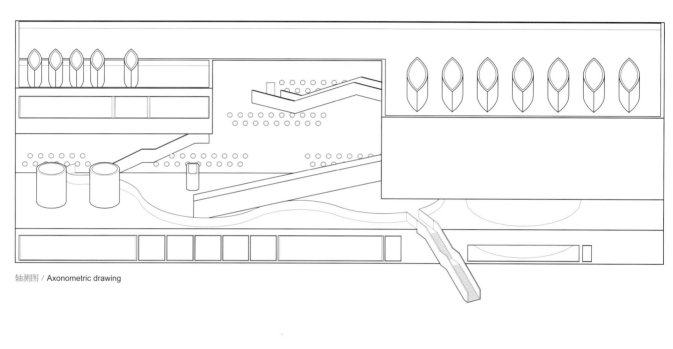

轴测图 / Axonometric drawing

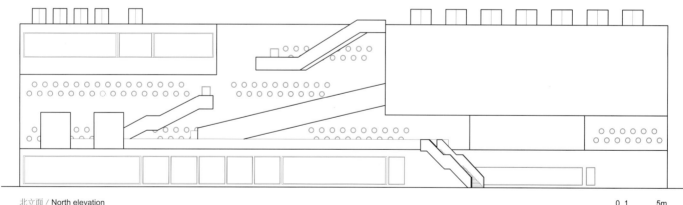

北立面 / North elevation

Sujiantuo Primary School is a community school in the north outskirts of Beijing. Total building area of the school is 4969m². Due to limited base, how to deal with the activity places is crucial in the design. The main building is on the south of the base and the playground is on the north. The layout of the playground is tilted due to the base condition. The main building is 74m long and 24m wide. The main classrooms are in the south of the building for enough light. The area of F1 is 1810m². The roof of F1 is an outdoor platform in the north and is connected to other floors by stairs. Students can easily reach here in the break where diversified activities could be carried out. It's connected to the gym classroom and playground by stairs. The platform is connected to F2 public classroom and F3 corridor too, and thus make it a public plaza. The lighting roof of the gym playroom and art classroom brings high-quality space feeling. The space flow line is simple and specific. Different space types are closely associated via proper space transformation. but not intrefere with each other.

1	后勤用房	Logistics room
2	职工食堂	Employee dining room
3	员工宿舍	Employee dormitories
4	教学用房	Teaching rooms
5	办公室	Office rooms
6	德育展览室	Merit education exhibition room
7	总务仓库	Logistics warehouse
8	会议接待室	Meeting & reception room
9	阅览室	Reading room
10	普通教室	Common classroom
11	多功能厅	Multi-function hall
12	体育活动室	Gym playroom

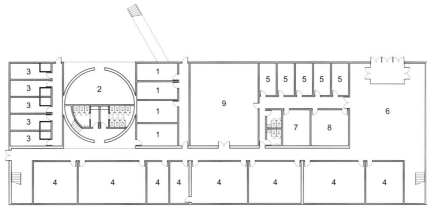

首层平面图 / The 1st floor plan

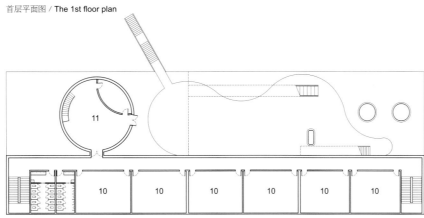

二层平面图 / The 2nd floor plan

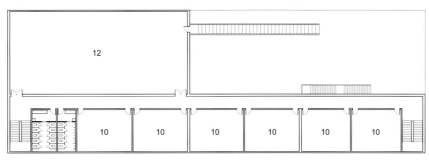

三层平面图 / The 3rd floor plan

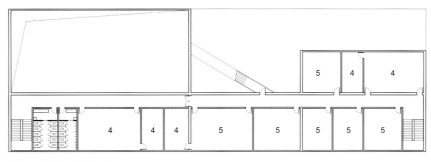

四层平面图 / The 4th floor plan

0 1 5m

私宅 杭州
Private Villa, Hangzhou, 2010

这是一个为私人设计的住宅，住宅的基地处于一个山脚下的缓坡地形上，整体趋势北高南低，基地西面和北面紧挨着一条进村的道路，南面是一条小河。为了避免这个低矮的地段给人一种处在低谷的感觉，设计时沿着基地北面较高的道路边上做了一条比较高的墙，南侧为了可以观赏到河塘良好的景观，在住宅前面的围墙做了开敞处理，出于安全的考虑，住宅由四面院墙围合出了一个100米×19米的长方形院子。住宅的总建筑面积为1900平方米，其中布置有7间卧室、1间书房、大小2个起居室、厨房和餐厅。由于西侧临进村的主要道路，因此将基地的西侧作为住宅的主入口并设置了3间车库。庭院中保留了场地内原有的一棵大树，并作为院内的主要景观，树边上的梯形的水塘，即作为景观的同时也是平时收集雨水的蓄水池。

住宅略高于庭院，目的是使人在室内也能够获得朝向河面的良好景观。沿河道一侧设置有两个次入口，靠西侧的是一条很长的坡道，人从坡道上漫步时，可以欣赏到河塘一侧的在不同标高上的景观变化。靠东侧的次要入口是一个台阶，进入院子后正对厨房，这里是搬运生活物资的通道。在厨房一侧设置有楼梯可直通屋顶，屋顶可作为晾晒农作物的作业平台。

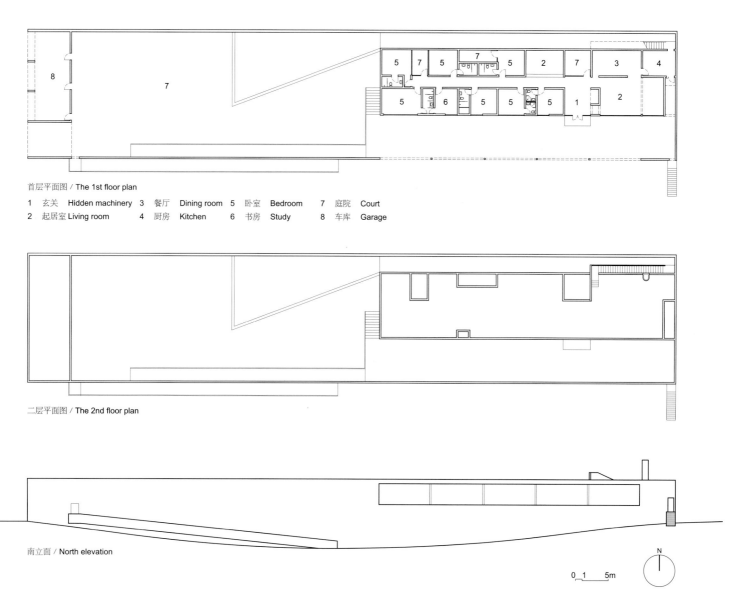

首层平面图 / The 1st floor plan
1 玄关 Hidden machinery   3 餐厅 Dining room   5 卧室 Bedroom   7 庭院 Court
2 起居室 Living room   4 厨房 Kitchen   6 书房 Study   8 车库 Garage

二层平面图 / The 2nd floor plan

南立面 / North elevation

This project is a private house located in a slow slope at the foot of a hill. On the whole, the north base is higher than the south. There is a road in the west and north and a brook in the south. To avoid the feeling of being at the bottom, a high wall is designed along the north road. To watch pond scenery in the south, the wall at the front is opened. A 100mx19m quadrate court is enclosed by wall for safty. Total building area of the house is 1900m². The west of the house is adjacent to the main road, and thus the main entrance is placed there with three garages. An old reserved tree in the court is the main scenery. Beside it, a steped pond is used as reservoir for rain water. The living part is higher than the court, so the residents could enjoy the river scenery. Two secondary entrances are beside the river. The west one is a long slope which enable people to experience scenery changes of the pond. The east one is a ladder, leading you into a court infront of a kitchen, which is used to convey living materials. A stair beside the kitchen leads to the roof where crops could be aired.

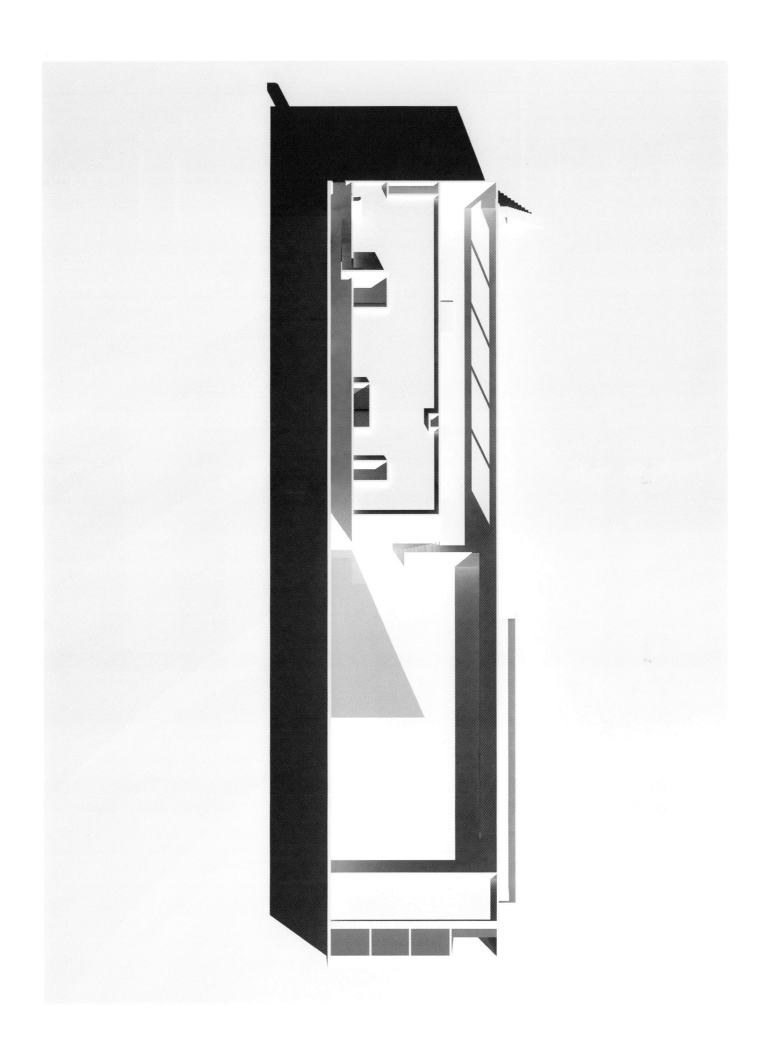

鄂尔多斯康巴什第六中学校设计 鄂尔多斯

Ordos Kangbashi NO.6 Middle School, ordos city, 2010

这是为鄂尔多斯康巴什新区设计的第六中学校舍。设计基地是一个南北向的长方形地块,教学楼放置在场地北侧,南侧放置一个400米操场,由于场地比较狭窄,可用作活动场地的空间有限,为了给学生们提供更多的活动场地,将北侧的教学楼一层架空,在一层仅放置室内体育活动场、音乐教室、报告厅等需要大空间的功能房间。教学楼的主体分为6排南朝向的体量,最北面是行政办公楼以及教室和学生的宿舍,南面5排布置为教学楼。在这5排体量中,三层部分全部布置为学生日常上课的普通教室,二层放置的是实验室、计算机房等功能性教室,这样的布局可以使每一个教室都能获得均等良好的采光条件。

6排建筑体块之间的东西两端,用两条走廊连接起来,体量之间的空隙作为建筑的中庭,即能够给教室提供采光,同时也可以在其间布置绿化。在这些空隙间设置室外平台,增加可用于室外活动的场地。

考虑到学校建筑不仅是作为学生学习知识的一个场所,同时更重要的是在设计时还注重了空间的丰富性处理,目的是使得孩童们在少年期间便生活在空间丰富的建筑环境之中。此外,运用了教室层高与宿舍层高不同的特点,将教师宿舍布置在五层。

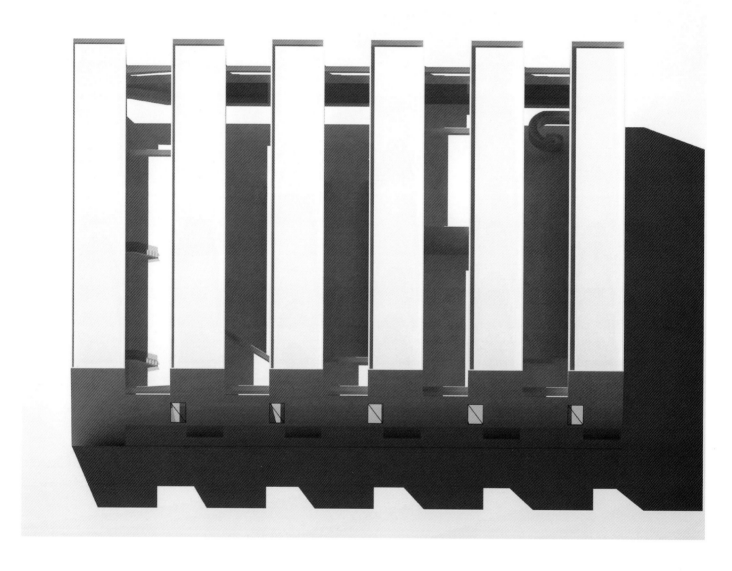

This design is for the 6th Middle School in the Kangbashi New Zone of Ordos city. The base is a quadrate land from south to north. The teaching building is put in the north and the 400m playground the south. The F1 of the teaching building is elevated to expand activity space which is limited due to the narrow site. Large function rooms such as indoor gym, music classroom and reporting hall are deployed at F1. The building is designed as six main bodies facing south. The administrative office and student dormitories are in the northest one. The teaching classrooms are in the five south bodies. F3 is designed for common classrooms and F2 the laboratory and computer classrooms. Such layout enables each classroom to get enough light.

Six bodies are connected by two corridors in the east and west. The gaps between the independent bodies are used as courtyard for light and greens. Outdoor platforms are inserted to the gap for activities.

Space variousness is focused in the design to provide children a environment of mutiple spaces.

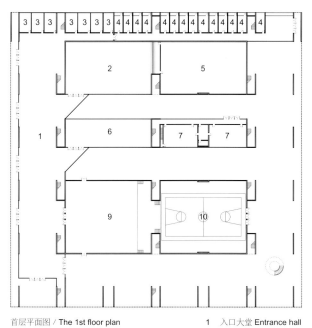

首层平面图 / The 1st floor plan

1 入口大堂 Entrance hall
2 食堂 Dining room
3 办公 Office
4 宿舍 Dormitory

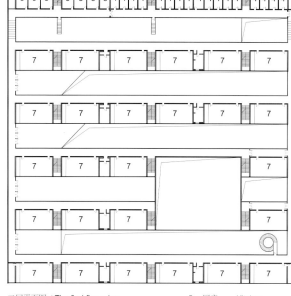

二层平面图 / The 2nd floor plan

5 厨房 Kitchen
6 阅览室 Reading room
7 实验教室 Laboratory room

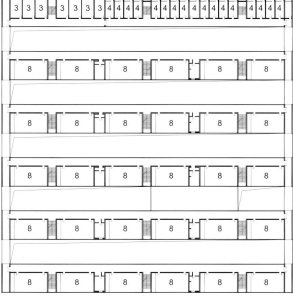

三层平面图 / The 3rd floor plan

8 普通教室 Common classroom
9 多功能教室 Multi-function classroom
10 体育活动室 Gym playroom

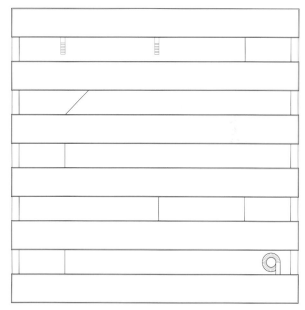

屋顶平面图 / Roof floor plan

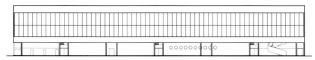

南立面 / South elevation

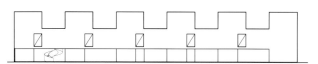

东立面 / East elevation

119

无锡天鹅湖会所A+B设计 无锡

Wuxi Swan Club A + B, Wuxi, 2007

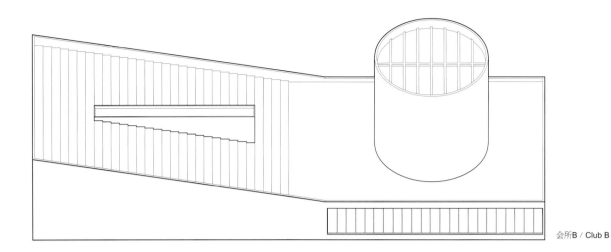

会所B / Club B

会所A+B拟建于无锡市天鹅湖住宅区的一个丁字路口,两个建筑隔马路相对,分别紧邻两个湖区,其中会所A的建筑面积为1000平方米。从功能分区上,业主希望将一层的开敞空间作为开敞的办公空间使用,同时希望内部空间可进行灵活分割和使用;二层作为财务室来使用;三层和四层作为总经理的办公空间。设计时在一层开敞空间的屋顶上设置一个开敞的屋顶花园,为办公人员提供一个休息的场所。会所B业主希望能够作为会议和餐饮使用,其建筑面积同样为1000平方米。在设计时,将一层大厅作为多功能场所,既是入口大厅,又可以作为举行某种仪式的场所。为此,在大堂的中心设置一个高耸空间,以增加这一空间的戏剧性效果。在一层还布置有可以作为多功能使用的会议及餐饮空间,厨房被安排在地下部分。在二层有一个中心通道可以直接通向屋顶平台,根据建筑的空间布局,屋顶做成一个阶梯状的小型屋顶剧场,由于建筑位于天鹅湖畔,在假日和怡人的气候时,在屋顶剧场上可以举行各种的露天音乐会。

The chamber A+B will be constructed at a T junction in the Wuxi Swan Cell. Two buildings are opposite on both sides of the road and are respectively adjacent to two lakes. The building area of the chamber A is 1000m². The owners hope the open space at F1 could be used as open office and could be divided flexibly. The F2 is used as financal room. The F3 and F4 are used as office space for general managers. An open roof garden is designed at the top of the open space at F1 for staffs. The owners hope to use the B building for meeting and dining. Its building area is also 1000m². F1 hall is used as the multi-function site which can be used as entrance hall and also a site for ceremony. A towery space is designed at the center of the hall to enhance dramatic effects. The multi-function meeting and dining space is placed at F1. The kitchen is under the ground. At F2 a center passage is connected to the roof platform which is designed as a theater with steps according to the space layout. Besides the lake, open concerts can be held at the roof theater on the holiday or in comfortable weather.

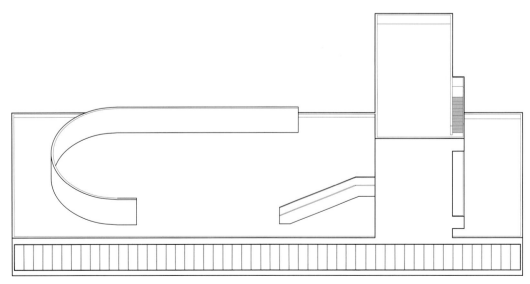

会所A / Club A

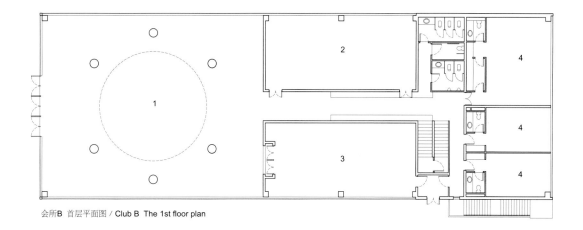

会所B 首层平面图 / Club B The 1st floor plan

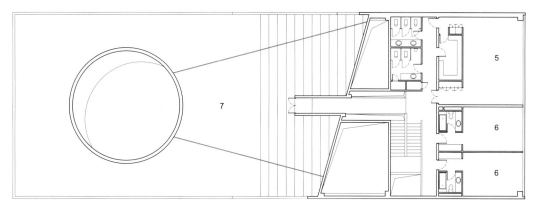

会所B 二层平面图 / Club B The 2nd floor plan

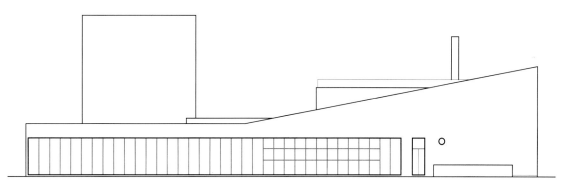

会所B 西立面 / Club B West elevation

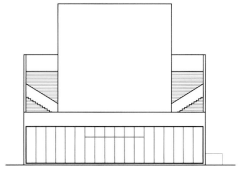

会所B 北立面 / Club B North elevation

1	门厅	Entrance hall
2	报告厅	Reporting hall
3	餐厅	Dining hall
4	包间	Loge
5	会议室	Meeting room
6	客房	Guest room
7	屋顶平台	Ceiling platform

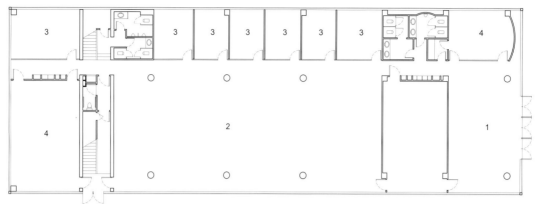

会所A 首层平面图 / Club A The 1st floor plan

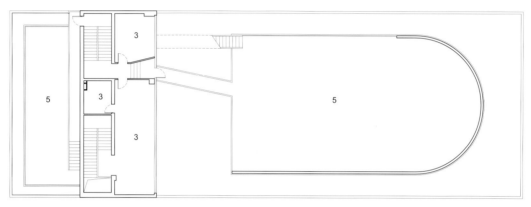

会所A 二层平面图 / Club A The 2nd floor plan

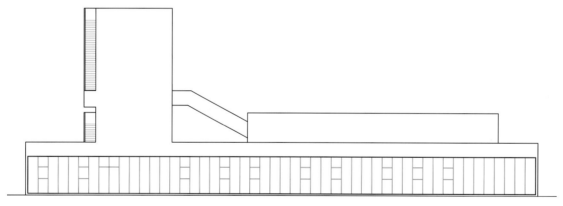

会所A 西立面 / Club A West elevation

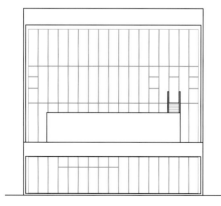

会所A 南立面 / Club A South elevation

1	门厅	Entrance hall
2	开敞办公区	Open office area
3	办公室	Office room
4	会议室	Meeting room
5	屋顶平台	Ceiling platform

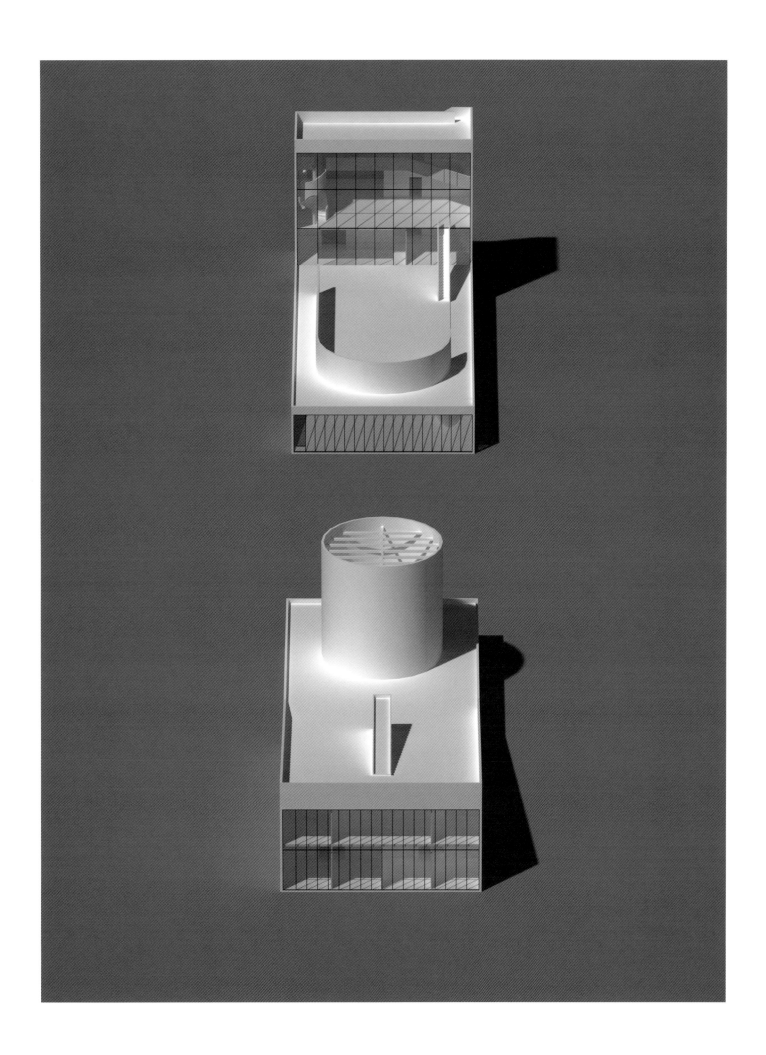

西溪学社 杭州
XiXi Learning Community, Hangzhou, 2007

西溪学社是西溪湿地第三期艺术村的12个项目中的一个局部项目。业主希望通过这样的项目开发，为文化和艺术界人士提供一个进行艺术创作和学术交流的场所。整个艺术村汇集了国内12位青年建筑师，每一位建筑师分别利用不同的地块设计不同使用目的的建筑。在本区域内，业主希望建一组叫西溪学社的文化交流和艺术创作场所。引发该提案的缘由是因为在杭州的西湖边有一个西泠印社，西泠印社曾经在中国近代文化发展上起到了很大的作用。鉴于这样的思考，业主希望在西溪湿地中同样建造一个能够传播中国当代文化和艺术的场所，不定期地请一些艺术家和学者来这里创作并为学生授课，同时也为艺术家本人提供一个怡人的创作环境。西溪学社的建设基地很特殊，用地分散且彼此间被水塘所分隔。业主希望整体的建筑面积能够控制在3800平方米左右，同时能够满足学社的使用需求。面对这样的用地状况，如何满足既有集中、又有分散的功能需求，同时兼顾创作空间、交流空间以及展览空间的结合是这个设计的关键。考虑到西溪湿地这样不可多得的良好自然景观以及湿地中树木的尺度，设计时将建筑的体量拆分，并尽可能地采用聚落的智慧是这个设计的主旨。由于西溪学社用地分散，设计中采用离散式聚落布局的方式，力图能够获得一个得体的处理结果。沿着这样的思路，设计时将西溪学社的整体布局成为一个离散式的学社建筑群。所谓离散，是指虽彼此分开，但彼此之间又有连带性的关系，类似于游牧性质的空间体系，而这一点事实上与当下的社会状态又有着明显的契合。仔细思考一下，当代由于信息技术的发展，交通的发达，人与人之间的物理距离和心理距离关系的差距越来越大，有时物理距离虽近但是彼此心理距离遥远，而物理距离遥远，但是通过现代电子网络联系却可以编织出一个超越物理距离的近距离网络。这种超越距离的当代社会的状况从真实物理距离的角度来

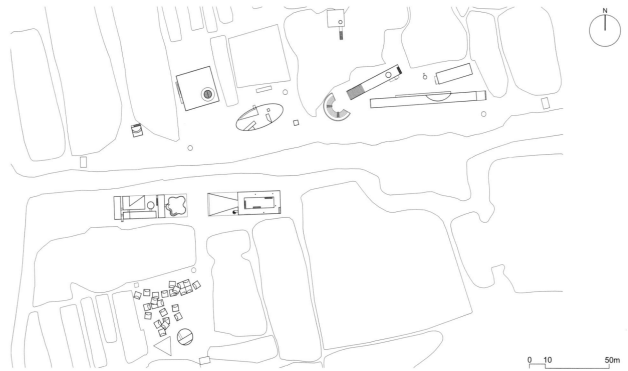

总平面图 / General plan

Xixi Learning Community is a local project in the phase III art village of Xixi swamp. The owners hope to provide a site for art invention and science exchange via this project. The art village attracts 12 young architects of China. Each architect designs buildings of different functions in different districts. Xixi Learning Community is originated due to Xiling Yin Community beside the Xi Lake in Hangzhou, which is eminent in the culture development of modern China. Considering this, the owners hope to construct a site to spread the contemporary culture and arts in Xixi swamp, and irregularly invite some artists and scholars to create and teach here. The construction base of Xixi Learning Community is distributed and separated by ponds. The owners hope to control the whole building area to be about 3800m² and satisfy the requirements. It is important to meet the centralized and distributed function requirements and combine the invention, communication and exhibition space in the design. Considering nice nature scenery and tree scale in the swamp, dimension split and cluster idea is formed. The type of discrete clustering layout of settlement is used due to distributed construction land. Based on this idea, we deployed the Xixi Learning Community as an architectural complex. The so-called discrete layout indicates seperations but also connections of space, like the space system for nomadism, which matches contemporary condition well. Nowadays, with development of information and traffic technology, physical and psychological distance between people are uncorrelated improvingly. People far from each other in distance can be closely associated via network but those who are near in physical distance might be remote phycologically. This contemporary society condition is discrete from the view of the physical distance, but the physical discrete

看，彼此之间是离散的，而这种物理离散状态表征的存在或许又构成当代的社会特征。这种离散的社会彼此之间得以存在的架构依靠的是信息的结合，从这个视点出发，我们可以把每个人看作是处于离散状态的点。从传统聚落的角度看聚集与集合向心是一个显著的特征，这种特征的主要目的是限于交通与信息不发达的状况，尽可能地缩短彼此间的物理距离。就是说过去的聚落是一个集合式的，大家必须要面对面才能沟通，而现在所有这些实际上可以通过电信网络来实现。通过现代的手段，能让物理层面上离散的状态事实上却彼此集中在一起。鉴于这样的理解，西溪学社便从这种当代的状态中获得表征。即表面上看来，学社的建筑彼此是很分离的，而以"共同幻想"作为逻辑表征，离散的聚落便成为整体。

西溪学社共有10个建筑和6个构筑物组成，总用地范围为5.98万平方米，其中的西溪学社的总建筑面积为3835平方米。一层的占地面积为2710平方米。整个湿地中建筑的体量采用明确的几何形体组合，以使其产生共同幻想特征。设计时在采用简洁几何体进行布局的同时，注重于聚落中所必需的"微差"的设置，使体验者能够在若即若离中会到一种丰富性的存在。鉴于基地中可以用于建设的场地面积较为分散，因此对建筑的体块进行切分是进行整体布局的手段。即将大而整的体量打散，形成离散式的空间布局，进而也使建筑与自然之间得以更紧密的融合。实际上在中国传统的绘画中，留白是非常重要的。白色不是一个虚无的状态，而是意境的生发之处。受这种在白纸上以墨作画的启发，设计时将15个建筑体量离散地布置于层次丰富的湿地环境之中的同时，将建筑做成白色，使其如同白色的宣纸，作为背景以衬托湿地美妙的自然景观构成的前景。让建筑作为背景而存在，让建筑融于画中，并与湿地的自然环境互相映衬，从而让人们体验到白色的空无之处生发出无限的妙境。

condition may represent the contemporary society feature: it depends on the connections of information. From this view, we can regard each person as a discrete point. Considering the traditional settlement, clustering and centering is a remarkable feature, which aims to shorten the physical distance as much as possible. It is caused by the underdeveloped traffic and information system. In another words, the used settlement is for assembled people to communicate in a face-to-face mode. These days, people who are remote from each other can converge via modern means. Based on this understanding, Xixi Learning Community obtain this feature from the contemporary culture. Seemingly, the Learning Community is separated from each other, but the discrete settlement is composed as a whole from the logic of "common consciousness".

Xixi Learning Community consists of 10 buildings and 6 components. The total area is 59800m^2 and the total building area is 3835m^2. Simple geometric objects are used in the buildings to generate a feature of "common consciousness". Simutaneously, "small differences" of the settlement is focused in the design, so people could take a journey full of experiences. Considering the discrete base, the building is divided to form a discrete space layout to fuse with nature. White is significant in traditional Chinese painting. It is not only a state of nihility but also where consciousness generates. From this idea, 15 buildings are discretely deployed in the swamp, painted white and worked as white rice paper which can highlight the wonderful swamp scenery. Existing as background, The buildings are fused in the painting and highlight each other with swamp environment. People can experience the "limitless wonderful environment brought by empty and virtual space" from the white.

鄂尔多斯康巴什住宅组团设计 鄂尔多斯

Ordos Kangbashi Residential District, ordos city, 2011

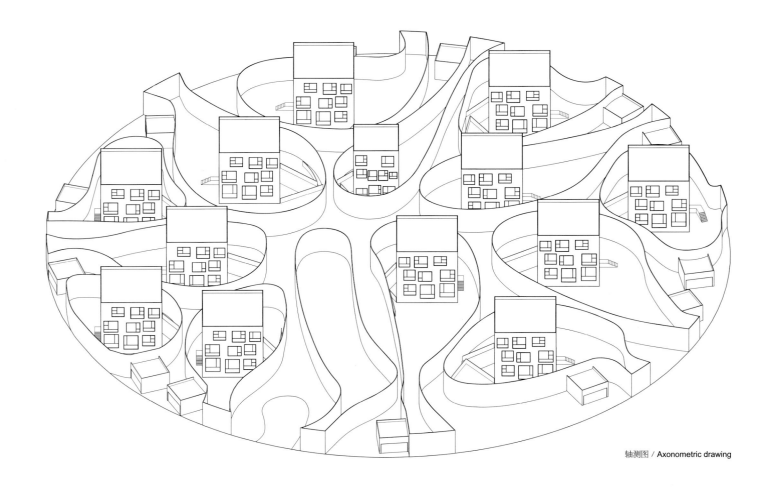

轴测图 / Axonometric drawing

这是一个在鄂尔多斯康巴什地区开发的住宅项目，该方案为开发区域内的一个小的组团。组团内要求建设5000平方米的建筑面积，其中要求每一户的基本户型控制在500平方米左右。设计时根据业主的要求，将每一栋住宅布置为10米见方的体块，每一栋建筑均为地上3层和地下1层所组成。同时在庭院中均布置有下沉的庭院，以使得地下室也能够拥有充足的采光。

事实上，就郊区的住宅而言，它不同于在城市中建房子，由于这个项目用地的容积率相对较低，如何在这里为使用者创造出多种使用可能性，为使用者提供一个丰富的空间就变得很重要。特别是在这个组团里，住居又基本都是500多平方米一栋的户型，这种户型不论在造型上还是面积使用上又都基本相近，如何在这种相似里面产生个性这个问题，在本方案的设计中就变得很重要了。住宅中的面积并不是单纯地为了满足居住这么简单的功能，而更重要的是如何让这个面积能够提供多种可能性。能够产生更多的生活乐趣，亦或是在空间当中造成一种空间印象，让人在里面生活起来能够获得丰富的感受。

在这个设计中，实际上重点还探索了墙的要素的应用问题，墙体这个要素虽然在平面上表现的仅仅是一条线，但是平面所表现的线实际上是和空间相对的。在建筑学里面，线本身不是单纯的图案的问题，它是一个空间问题的同时，也标示着人在里面的行为问题，实质上每一条线的划定，都规范了人在空间里面行走的方式和人的行为的逻辑性。当然就墙体本身而言，有两个意义存在着：一是主观上从空间角度进行思考的墙体，另外一个是为了一个立面或者是为了围合一个范围所形成的墙体，而这两个均称为墙体的对象物，本质上是不一样的。从空间的角度，或者说是从人的行为的角度来思考空间的布局，任何的一个线的变化和人的内心的感受是能够获得互动并具有一致性的。这个住宅组团的设计，采用沿外围布置车道的做法，从周边能够直接进入车库，在圆形的基地内部，是一个充满曲折律动的人行街道空间。伴随墙面的曲动，空间的张与缩紧密地呼应和结合。在该区域的一侧，还设置有一个由围墙所围合的街心花园，这个围合的空间，平日是一个小的园林，但在举办活动时，这里又是一个聚会的露天会所。

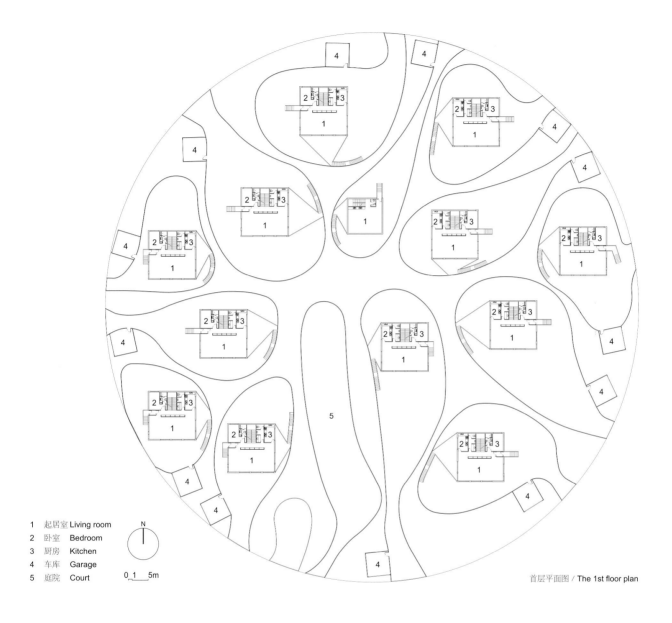

1 起居室 Living room
2 卧室 Bedroom
3 厨房 Kitchen
4 车库 Garage
5 庭院 Court

首层平面图 / The 1st floor plan

This project is for a housing group includes 5000m² building area in the Kangbashi area of Ordos city. The basic area of each house is controlled to be about 500m². Each house is deployed as about 10m block according to the requirements of the owner. Each building consists of three aboveground floors and one underground floor. The court is sunken to make sure the basement has enough light.

In fact, houses in outskirts are different from them in urban areas. Because the plot ratio of this project's construction land is relatively low, it is significant to create multiple space possibilities. Especially when the basic area of each house is about 500m² and the houses are similar in style and function, it is crucial to generate personality in similar space. The space of the house is not only for simple utilization, but also to provide multiple possibilities, generate more living interests or different space impressions in the space. Actually, this design focuses on applications of wall elements. Although the wall element is only a line in the plane, it relates to the real space. In architecture, line is never a pure pattern problem, but a space problem. Simultaneously, it symbolizes the human behaviors in the house. In fact, each line division standardizes the walking mode of the humans in the space and logic of the human behaviors. The wall exists with two meanings. One is the wall which subjectively thinks from the space view. Another is the wall which is for a vertical face or enclosing a range. They are both called the object of the wall but are different in essence. When thinking space layout from the view of the space or human behaviors, any line change interacts with and is consistent with the feeling of the human heart. The road is deployed along the periphery in the group design, and the car can directly drive into the garage from the periphery. The devious and wonderful pedestrian street is inside the circle base. With devious motion of the wall, the space tension is closely associated with space shrinkage.

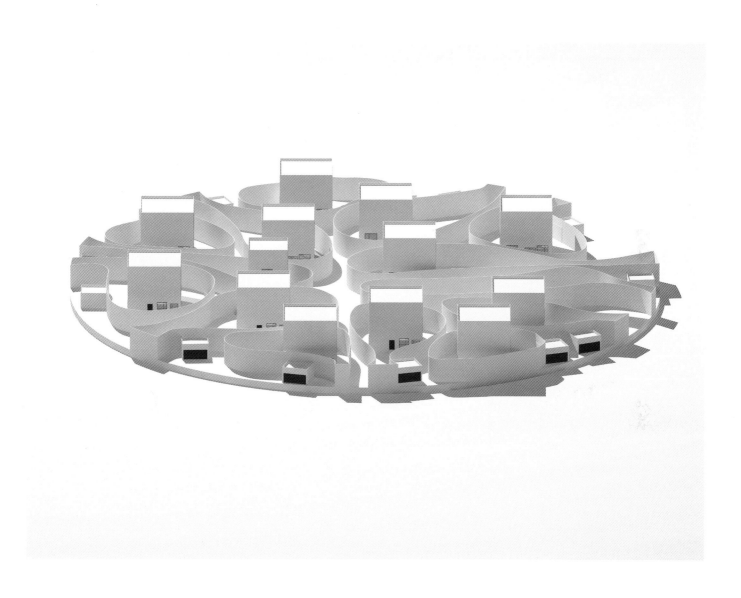

王昀——几何抽象与冥想　Wang Yun—Geometric Abstraction and Meditation

文/方振宁（建筑评论家）
By Fang Zhenning (architecture critic)

王昀有一点和其他中国建筑师不同，那就是他以一名建筑师出现的时候，实际上他有着学者的背景。换个角度我们也可以这样看，当他发表自己的学术观点和著作时，他是从一位建筑师的立场出发的。无论是在欧洲还是在日本，这种现象不足为奇，但是在中国却是个例。

王昀的设计是有逻辑而严谨的，那是因为他生来喜欢理性的东西，理性是经过提炼的结果。提炼，就是善于将事物抽象化之后再表达。如果没有这种能力，人类的文明就不会有今天这个样子。在两河流域就不可能有楔形文字出现，也不可能有金字塔；在欧洲就没有希腊的数学和科学思想的诞生；在中国就不可能有象形文字，当然也就没有勾股定理和易经。如果没有抽象的能力，人类就产生不了数的概念和黄金比，没有二进位制怎么能有电脑的问世呢？所以说，如果劳动创造了人的话，那么是由于有了抽象的能力才使人类进入到高级阶段。

王昀的设计生涯应该从他2001年从日本回国之后的第一个设计"60平方米极小城市"算起，这一设计是对城市组织结构的抽象描述。而那些几何形体，比如：圆锥、立方体、长方体、斗形等都是抽象出来的符号。十年了，这些符号一直在各个设计中不断以新的姿态出现，从原来静止的，可以平行的移动状态，到相互穿插飞扬的碎片状，都统一贯穿在一种氛围里。集大成的作品是浙江西溪湿地的项目离散的聚落，它几乎汇集了过去十年来，他在几何抽象设计方面的各种语言。

其实要全面了解王昀的设计，还需要把他出版的书拿来作为辅助读物，书里清晰的记录了几何抽象体块，是如何像零部件一样的被组织成一个整体，那过程就像组装高达一样有趣。

王昀的几何抽象设计，让我想起艺术史上两位重要的艺术家丢勒（Albrecht Dürer，1471～1528，德国）和莫兰迪（Giorgio Morandi，1890～1964，意大利）。丢勒的版画作品《忧郁1（Melencolia1）》（1514年）中那些神秘的几何形体和各种古怪、诡异、源自想象的东西启发人去思考。甚至太阳都被画成三角形，作品中多种暗示性，让它成为美术史上的一个至今没有揭开的迷。

而莫兰迪被誉为冥想的画家，他的绘画是通过对普通和平凡的物体的描绘，寻找和宇宙相连的要素。寡言的莫兰迪在1957年说："我记得伽利略说过，真的哲学、自然的书不属于我们那种用阿拉伯字母的文字写的书。而自然的文字是三角、四角、圆、球、锥形等几何形状。"原来这些是莫兰迪绘画的哲学依据，是不是可以说这也是王昀建筑设计的哲学观？

Wang Yun is different from other Chinese architects in one respect. He is an architect with scholarly background, or we can say that he delivers his academic viewpoints with the starting point of an architect. This phenomenon is quite common in Europe and Japan but not in China.

Wang Yun's preference to rational things which abstracted from specific ones lead to logical and rigorous aspects in his works. Without the ability of abstraction, human civilization would not be what it looks like today. Arrowheaded characters and pyramid couldn't appear in Mesopotamian; Greek mathematics and scientific thinking couldn't appear in Europe; hieroglyphics, Pythagorean proposition and Yijing couldn't appear in China, not to mention the concept of numbers, golden ratio, the binary scale or even computers. If it is labor that creates human, then it is the ability of abstraction that brings humans into the advanced stage.

Wang Yun's design career began from his "60m2 smart city" which was complicated in 2001 after his return from Japan. This project was an abstract description of city organizational structure: cone, cube, box and bucket were all abstracted symbols. During the next 10 years, these symbols have been appearing in all his works with various forms: from stable one to moving one and then to flying chips, but all of the states share the same atmosphere. The most important work of Wang Yun was separated settlements in Xixi Wetland in Zhejiang, which is a collection of the architect's design languages during the past 10 years.

Actually, Wang Yun's publications can help us understand his designs better. The process of organizing all the geometric blocks into a whole was recorded in the books. The process is as interesting as assembling the robot Gundam.

Wang Yun's geometric abstraction reminds me of two important artists in art history. Dürer (Albrecht Dürer, 1471-1528, Germany) and Morandi (Giorgio Morandi, 1890-1964, Italy). Dürer's prints "depression 1 (Melencolia1)" (1514) inspire our deep thoughts through mysterious geometric shapes and a variety of weird, strange things of imagination. Even the sun was drawn as a triangle. Multi-suggestions made the work an art mystery. Morandi was known as a meditation painter, his paintings tried to find out connections between the world and the universe by depicting ordinary objects. Reticent Morandi said in 1957, "Galileo had said that true philosophy was not written in Arabic alphabet but in geometry including triangle, rectangle, round , ball and cone." This is the philosophy basis of Morandi's paintings, is it also Yun Wang's architectural design philosophy?

对谈 / Discussion

柳青 + 黄居正 + 王昀

情感、记忆与历史意识

Emotion, Memory and the Sense of History

柳青
UED杂志执行主编
黄居正
建筑评论家
建筑师杂志社主编

第一部分：历史观

后现代主义——虚无的文脉

柳青：我看过您的不少作品，很明显都具有您的风格，这些是不是与您的经历有关系呢？

王昀：我们上学的时候正是20世纪80年代，流行的主要是欧美风格、后现代主义。记得当时理论家们评价现代建筑大多是"腐朽的"，没有人性、缺乏人情味的。希望要做"温暖的"建筑，并且要跟机械的、没有人情味的现代建筑决裂。所以当时我们认为现代主义已经过时了，真正的流行的是后现代主义，建筑已经从现代主义到后现代主义了。20世纪90年代初我到日本留学的时候，应当是日本后现代主义最繁荣的时候。我接触的一批建筑师，比如最典型的是黑川纪章和矶崎新，矶崎新当时是后现代主义的旗手，到接下来的六角鬼丈、毛纲毅旷，还有一大批的日本建筑师，在日本有很多的实践机会，他们能够将方案实现，所以后现代理论大量的实践是在日本。

1993或1994年的时候，有机会去欧洲看现代建筑。我们到了巴黎后去看了萨伏伊别墅——勒·柯布西耶作品，那曾经是后现代主义集中批判的一个建筑，批判它没有人性、很机械。但是看了以后你会发现萨沃伊别墅，根本就不是照片里看的那种冰冷的感觉，而是一个非常温暖的建筑。很多照片我们看上去很冷的建筑实际上很温暖，而很多看上去很温暖的建筑，实际上很冰冷。为什么会这样？就是由于后现代主义建筑是从虚无的文脉和造型上找东西，而现代主义建筑的本质就是从空间的内涵里面对建筑进行叙述。

现代主义——空间的艺术

黄居正：现代建筑在日本起步比较早，在20世纪二三十年代，就与西方现代建筑亦步亦趋了。到了50年代的时候，日本许多的建筑师，例如丹下健三、文彦、吉阪隆正等都是在模仿勒·柯布西耶，在日本国内建成了一批柯布风的清水混凝土建筑。因为当时勒·柯布西耶在印度做了昌迪加尔的规划，还建了三个纪念碑式的建筑。日本人觉得现代主义建筑也能和地域风土，尤其是东方文化相结合。到了后现代主义建筑时期，80年代的时候，在日本筑波看到了矶崎新的作品——筑波中心大厦，激动得不得了啊！当然，在所有的后现代的作品中，我认为矶崎新的作品应该算是最成功的。同时，后现代对中国来讲还有另外一层含义，因为内忧外患，中国和现代主义建筑总是阴差阳错，没有跟上现代建筑的步伐。近代以来民众潜藏的民族情感一再被激发，国家主义、民族主义、民粹主义一抬头，后现代主义建筑便可与中国当时的历史情景和文化心理若合符节了。所以后现代主义在中国甫一出现，现代主义建筑就被弃若敝屣，建筑师都奔着后现代去了。那王昀到日本是在20世纪80年代后期90年代前期，正是日本后现代主义非常兴盛的时期。这与日本的经济状况有关联。当时日本处于泡沫经济时期，经济非常繁荣，到了1993～1994年，泡沫经济破灭，后现代主义也就随之烟消云散了。记得1993年隈研吾做的M2大厦么（马自达公司办公楼）？那是他的成名作。那是让人看了特别惊讶的建筑，抛弃了一切形式美的原则，丑陋、傲慢，但居然能盖出来，可想而知那是一个怎样烧包、疯狂的年代啊。当然，日本的后现代跟美国的、欧洲的，还是有一些细微差别的，毕竟日本的文化中没有连贯的西方历史文脉，日本建筑师是把美国的、欧洲的东西拿过来进行拼凑，形成他们所谓的后现代。

后现代主义破灭以后，日本比较有代表性的建筑师就是伊东丰雄。伊东丰雄实际上是批判后现代主义的，但是他对日本的社会、经济、文化状况有较好的理解和把握。他首先回归到了现代主义建筑的原点，把那儿作为一个切入点，一个出发点，一步步地发现与日本当代社会文化的结合点，寻找出自己的风格。他的一些作品包括他早期的作品，和勒·柯布西耶等现代主义建筑师的作品之间有很多的相关性。他1976年的一个作品——中野本町之家（White U），形体简洁，内部强调空间抽象性的表达，可以看出受早期现代主义建筑影响的蛛丝马迹。

王昀到了日本以后，就像是考古人员一样仔细寻找，他发现各个时期、各个历史层面的建筑都有。经过研究之后，他认为我们的建筑要回到建筑的原点，他在中国所做的设计中可以发现这一点。后来他讲的游历，我觉得非常有意思。一个伟大建筑师的成长缺失不了建筑游历这一环，无论是作为建筑教育还是自我修为的一部分，譬如远的可以追溯到18世纪英国的大旅游（Grand Travel），法国巴黎美院的罗马大奖；近的有为后来者津津乐道的柯布的东方之旅，安藤忠雄出道前的欧洲建筑漫游。建筑是门空间的艺术，需要亲身的体验。某些作品，比如让·努维尔的Quail Branly博物馆（Musee du Quai Branly）做得很漂亮，各种颜色、各种材料都有，但是进入内部则什么都没有，很无趣。不体验，光看照片则无从知之。后现代就是一个布景性的东西，没有很深的意义的东西在里面，当代的一些建筑师多是这样的作品。但是如果去看勒·柯布西耶的作品，不管是萨伏伊别墅还是马赛公寓，空间都很丰富舒适，材料运用得也很得当。我想这点对王昀的创作肯定有很大的触动，我们看他的早期作品，像"60平方米极小城市"，与勒·柯布西耶的风格有些相似。受他的影响，这是免不了的，中国建筑师也不用避讳去学习别人的东西。因为实际上像王昀这一代建筑师，实际上是重新开始，没有从前人（中国建筑师）那里获得太多的东西。现代主义建筑是一个划时代的东西，在西方来讲，和古典式的建筑完全不同，具有颠覆性。我们中国的建筑师，没有从现代主义那儿学到一些基本的原则方法，反而是蜂拥而上，直接进入后现代，那是不可能做出任何好的作品的。所以王昀在欧洲的游历，对于一个优秀的建筑师来说是非常重要的。他看了很多的聚落，体验到了很多不同的形式构成、不同的空间模式、不同的材料表达。虽然他的作品有些地方似乎有勒·柯布西耶的影子，可是你进入其中，仔细观察，还是有很大的差异，有很多他在游历中积聚的、独特的感性成分在里面。

现代主义的建立——手工业审美体系的归零

王昀： 我觉得现代主义的产生与社会工业化是紧密相连的。欧洲从手工业到工业化的过程，经历了非常长的历史时期，而建筑从空间艺术变成了视觉艺术，是在19世纪达到顶峰。17、18世纪的巴洛克、洛可可风格是建筑在手工业社会的极致。人类从原始社会开始审美和技术的积累，花费了上千年时间，形成了一个完美的，社会、人文、思想等各方面都达顶点的手工业审美体系。而在20世纪初，这个体系被工业化生产，也就是机器完全摧毁，上千年的审美和对社会的理解都因为机械生产而归零。在这个现代化的、全新的社会，在一个瞬间归零的时期，我们要学新的语言、新的理解，这就是为什么建筑在以前归于装饰，直到19世纪末的时候才有人真正提出建筑是空间的艺术，而不仅仅是造型的艺术。这样的新体系实际上就把原来建筑的物质性给归零了，或者说建筑根本就不是物质，建筑是空间的，是森佩尔（Gottfried Semper，1803~1879）提出的这一理论。其后的现代主义建筑师试图在理解这个理论之后，建立起一个新的语言体系。这个语言体系实际是在20世纪20年代左右基本上就定型、就完成了。这是一个不同于19世纪之前的，全新的语言体系。这个语言本来发展得很好，但是其后的世界大战使其发展产生了停滞。

黄居正： 这个事情可能跟美国占主导地位有关系。实际上我们所批判的现代主义建筑，被很多评论家将矛头搞错了。他们认为冷冰冰的、没有人情味的，一部分是希契柯克和菲利普·约翰逊1932年建构起的国际式风格，一部分是格罗皮乌斯带到美国的异化了的包豪斯风格，而不是欧洲丰富的现代主义建筑实践。

王昀： 这其实讲的是建筑最本质的东西——空间，空间探讨如何组织房子与房子之间的关系，只有空间才能体现人的智慧，体现人与人的关系。建筑材料是木头或石头，从空间来讲是没有优劣的。我认为归零的这件事的意义就在这。钢结构、混凝土的房子与木结构的肯定不一样，但在空间的处理上是完全一样的。现代的房子实际上依然遵循现代主义建筑的空间关系。但是感觉会因为材料不同而有区别。如果专注于这一点而不是从空间的角度来挖掘空间内涵关系的话，现代建筑就成了一种形式主义了。当然结构也会对空间产生影响。

第二部分：王昀的建筑

建筑语言——心灵的感动

黄居正： 一般两种事物可以影响建筑的发展，一个是空间，另一个是技术。19世纪末法国结构理性主义者奥古斯特·肖瓦西的《建筑史》一书认为结构技术影响了建筑的历史发展，而现代主义建筑师则认为是空间。

从现代建筑角度来讲，建筑语言非常重要。在20世纪以前，在西方语言是思想的附属品，是从属于思想的。从哲学家维特根斯坦（Ludwig Wittgenstein，1889~1951)开始，语言成为一种独立的东西。在20世纪，现代主义建筑发现了与古典语言完全不同的现代语言，例如柯布西耶、赖特、密斯，他们共同建立了现代建筑的语言体系。其中，柯布西耶起到了非常关键的作用，他的两个原型，最重要的就是1914年的多米诺体系，在当时是革命性的，我们现在的几乎所有房子都是按照这个体系来盖的。对于建筑师而言，创作一个好的作品如果没有自己的语言在里面，也是很难获得认可的。在建筑里面最重要的，除了物质还有人的感情，诗一般的感情。营造诗性的空间氛围，激发调动人的情感，对于一个好建筑来说是不可或缺的。这种诗性情感不仅仅在现代主义建筑那里可以发现，在各个时期优秀的建筑中都能被发现、得到体验，古今中外，概莫能外。一个优秀的建筑可能存在着许多种价值，有艺术价值、历史价值、经济价值，但唯有情感价值最令体验者怦然心动。情感价值的塑造却是通过空间的组合和构成，和空间有很大的关联性。什么样的空间让你一进去就有一种感动？哥特教堂是以技术理性为主还是以空间为主？两种观点可能都站住脚，也可能都站不住脚。拱券技术和飞扶壁的不断改进，塑造了神圣的空间感；反之，从意志论的角度，有人说为了创造神圣空间而发展了哥特的技术理性也未尝不可。中世纪在欧洲历史上通常称之为信仰时代，那时候人们普遍虔诚地信仰上帝，需要从神那里得到感动、获得力量，创造与神沟通、感受圣灵的空间，是建造者唯一的希望。

在王昀的作品里面漫游，也许是因为设置了许多空间节点，在赋予趣味性的同时，也蕴含了种种情感的氛围。刚才说，在看Quai Branly博物馆时外观很花哨、很漂亮，但却没法让人感动。感动可以有不同，拿柯布西耶和密斯比较，在印度，柯布西耶的建筑给人一种震撼；看密斯的柏林新国家美术馆、芝加哥的伊利诺工学院建筑系馆，很冷峻，不会一下子打动你，但是细细地品味它的细部也会给你这个感动。都是伟大的建筑师，他和柯布西耶相比就是差给人一种扑面而来的感动。

空间语言——思想的传递

王昀： 建筑师和文学家有点相似，文学家使用文字来创作，他的目的是要告诉你一个故事，对你产生影响，让你有所体会；建筑师也有一种语言——空间性的语言，是建筑师的看家本事。每一个建筑师都对语言有自己的理解，你用什么样的语言来展现给别人、讲一个动人的故事，有很多种方式，后现代主义就是一种方式。现代语言，是手工业积累了那么多年的，完全从原点出来的东西。中国几千年的手工业的传统，完全归零，是一个全新的开始。

对我而言，个人的生活经历会对设计产生影响，可能会有相同的场景出现，但是采用的方式不一样。比如曾经生活过的学校，参观过的西安的窑洞，空间都很有意思。建好一个建筑之后我一般不太愿意去看，但是在经过很长一段时间之后再去看的时候，在里面走着，会想起自己曾经去过的地方、见过类似的空间。我们不知道别人的脑子里在想什么，脑子是一个封闭的世界，外界是看不见的，但是做了很多建筑以后，设计师在想什么，就好像小说家，一个作家，创作了很多作品之后，一看他的小说就知道这个人了。这是建筑真正有意思的地方。对我来说，看我的建筑，能够发现不同的时期思考的不同，潜意识里想的都在设计里展现出来，然后传递给别人，我对这个的乐趣远远大于房子造型好不好看的问题。参观者看过建筑之后，理解了我所想的，通过物质的展现、发生并完成了思想从设计者到使用者、观者、评判者、批评者之间的传递。设计里少不了造型的因素，但就我个人，我不希望让其他的要素干扰这种思想的传递，这是意境的传达。

我最近在做一本书《从风景到风景》，其实风景就存在于人的大脑里，就像作家是在写脑子里的世界，我用建筑的各种元素，将风景投射到现实社会中，唤起使用者脑子里的另外一种东西。有的房子看上去很好，可没有感觉，有的房子看了之后有点感觉，这就可以了。你的感觉可能就是我要传达给你的东西。我不觉得建筑师一个人就能改变这个世界，如果每一个建筑师都能把自己的失去的东西找回来，这个世界就会变得很美好。我不要我的建筑成为大众情人、人人都喜欢，我希望能有喜欢的人去喜欢我的建筑，能够理解建筑的人去理解就可以了。我不希望所有的人都做一样的东西，但是我希望的就是中国有建筑师做这种可以传达思想的建筑，这个是很重要的。我不想这个事要多特别，只要有人喜欢，这样就有这个建筑存在的价值，如果进去了之后再有所感悟的话就非常的满足了。

柳青：这就是一种心灵上的感应，这个建筑的使用者在跟你进行某种形式的对话。我突然间想起一个故事来，一个画家画了一只快要死了的鱼，画的含义没有人明白，所以一直没有人买。但是有一天一名企业家突然发现，画的意境就是自己在创业期间的写照，于是他出高价买了这幅画。这就是一种心灵的对话。

黄居正：刚才我讲的真正好的建筑，重要的就是能传递给你一种感动。你的记忆把以前所经历过的信息提炼成一个形象，传达给大家变成一个共享的记忆，这个对我们的历史的传承实际上是非常重要的。我举一个例子，罗马的河、广场、街道，去过，在那儿生活过的人都有记忆，每一个人的记忆汇合，交集部分形成共享记忆，通过历史书写、电影、文学传递给后人。你去看的时候，某种程度上都带有共享记忆的印记，之后经过你的观察沉思，如果足够敏感的话，部分共享记忆又会转变成你自己独特的记忆，如此生生不息。倘若建筑师能把个人记忆转化为他人能读懂的共享记忆，那么生活于其中的人，便能体验并为之感动。

聚落研究——封闭的共同幻想

王昀：现代我们对建筑的传承和理解不再是一个视觉符号问题，因为我们现在的生活，包括我们的记忆、经历，不再是地域性的封闭环境。对聚落而言，开始时一定是封闭的，但当开放了以后就会变质。一个村子交通不发达，没有任何外来的干扰，它保存的记忆一定是独特的记忆，当地人的记忆已经完全融在房子里头。当外来者进去的时候，会发现那里有很多对你来说是失去的东西，会有一种感动——或者是不太常发现的，或者是人性中最本质的东西。但是一旦这个村子与外界接触多了，这个记忆就完全被破坏了。我们在聚落研究的时候，有一个概念叫"聚落幻想"，就是这个村子里头有"共同幻想"，是相对独立的。但当外来的人进去以后，或者里面的人出来以后，这个"共同幻想"就打破了。当今世界的状况是所有的边界已经完全打破了，造成了记忆的模糊和错乱。在这样的情况下你的脑子不能像一个镜子一样只发生折射和反射，你的脑子应该是一个搅拌机。就是说进来的可能是苹果、香蕉、凤梨、桃子，经过脑子的搅拌变成一杯果汁，味道会完全不一样，不可能还是香蕉或苹果。现在这个世界的记忆范围与以往相比已经非常不同了，每个人都要唤起别人的注意，每个人用语言表达出来都是不一样的。我认为我现在在使用的语言表达我的记忆、我的情感、我的思想是没有障碍的。

传统与现代的融合——八宝文化

柳青：您的语言的表达上还是比较一致的。您一直在说中国的其他建筑师也是有各种各样的风格的，您的评价我记忆最深的就是"八宝粥"。

王昀：我们现在的文化的确就是"八宝文化"。现在的文化是非常丰富的，但是也是需要反思的。在这碗"八宝粥"里你可能一个很小的组成部分，是一个大枣或是一个花生，而不能把自己当成一碗"八宝粥"。

黄居正：传统的民居聚落文化是几百年凝聚起来的观念或者思想的体现，在我们看来是非常的美，但是一经过开发，马上就变"八宝粥"了。人都有一种矛盾——故乡和异乡。回到故乡，也就是聚落，就找到了一个可以回忆的地方；可是农村的人要走出来，他们向往别的文化、别的景观。作为建筑师就要调节故乡和异乡的情感，创造不仅能容下两种情感，并得到融合升华的空间，说白了就是传统建筑和现代建筑之间的矛盾。当然说说容易，做起来难。

王昀：对故乡的感觉，最好是留在记忆里面。记忆是最美好的，有的时候为寻找自己的记忆，回到故乡的时候，肯定是记忆破灭了。因为记忆是最好的，一旦改变了就不会认同了。对于建筑师来说，你的作品能调动别人的记忆，这个就是成功的。建筑需要一个抽象的、没有性格和倾向性的物体，来换取更多人的文化认同，这是有局限的，与诗和散文不同，一个抽象性的语言可能就会让你的感受越丰富，想象越发散，因此想象发散的空间就会越大。

柳青：这个是很难把握的。复杂的画面，可以精益求精的地画，但是画面越简单，就要求功底越深厚。

完全的抽象——人的需求丈量建筑

黄居正：绘画中几何抽象（还有抽象表现主义，以康定斯基为代表）在荷兰风格派那里表现得最为明显，也最为彻底，而立体派则是采用了具象手法。柯布西耶更多地吸收了立体主义的空间概念，当然不可避免地也受到了风格派的巨大影响。但是如果把它放在一个特定的区域里面，像20世纪50年代的日本，战后现代主义建筑重新开始，建了很多很有名的现代建筑，如筱原一男的作品，表现的是对日本传统文化的批判性理解。他是怎样做的呢？他把日本传统建筑的柱子放在房屋的中间，把它抽象化，变成一种象征性的形象。对抽象怎么理解？从哪个方面理解？我们的传统文化如何抽象？这个抽象有两种方法，风格派将宇宙抽象成横竖的直线，三原色；立体派从具象的角度出发，通过分解、综合物体，最终走向的是抽象性的空间形态，走的是不同的方向。具体就建筑说来，中国幅员广阔，东西南北地方差异也很大，在中国的某个特定的环境里面你怎么表现抽象性？

王昀：我个人认为抽象是在不断变化的。从纯粹的角度来说风格派的抽象还是有道理的。因为这种抽象的出发点是人，排除了所谓民族性的概念。以人为基本的原点，对我们今天谈的是很有意义的，现代世界越来越从人本身的角度思考问题。从人类最长远的大的目标来说，所有的问题都会最终回到人的原点上来。比如聚落，虽然处于不同的地域，但从整体上来看，不能说某个民族的建筑很差，只是在遇到问题的时候有不同的解决方式。例如日本的老房子，门高1.8米，这是按照日本民族的尺度标准；现在门的尺寸做得大了，这是因为按照人的尺寸来做了，还原到人的最本质的东西来判断。

黄居正：抽掉历史的、文化的意义，只考虑人的需要，就是用人的需求来丈量屋子的尺度。

王昀：其实在设计时，房子是按照设计师的理解来盖的，但是别人来看的时候，房子却唤起了别人的记忆。100个人来看了设计师设计的房子，他们肯定都有自己的理解，这种理解的丰富性远远大于建筑师的设想，这其实也更有意义。

黄居正：这有点超现实的意思。

王昀：房子本身就是超现实的。房子盖之前是不存在的，但是在盖起来的那一瞬间就变成了现实了。我希望的是房子这个对象物的本身，唤起的不仅仅是其本身，而是你看过这个对象物之后你能够唤起你的感受。

白色建筑——童年的记忆

柳青：你为什么喜欢白色呢？

王昀：我原来是在哈尔滨出生，在北方生活，哈尔滨这个城市具有很浓厚的俄罗斯风格，在我的记忆中：冬天的时候，经常一觉醒来整个世界就不一样了，到处是白颜色的，世界一下子就变成另一种形态了。头一天晚上和第二天早晨的对比就是一个童话，一个超现实对比。白色是很奇妙的颜色，不同的阳光、不同的时间段会形成不同的效果。后来来到北京，最喜欢的就是北海，蓝天，白塔。白塔就像是蓝天上的一个剪影，就像是天空被挖出一块白塔形状的轮廓，通过那个轮廓你似乎可以看到天空里面的情景。如果说我们的城市是一个现实的世界的话，我想能不能在现实的世界上面挖一个窟窿，设计一个非现实来，而现实中的白色就是为了制造这个窟窿，在现实中进行切割。

影像建筑——空间的投影

柳青：我觉得你是一个充满幻想的一个人，现在很多人都是在讲建筑和影像是什么，影像会给你的创作带来灵感吗？

王昀：我认为电影是不同于小说的那种叙事手法，在字里行间、在你的眼前造成的影像。我看一本书可能会花费几个星期，但是电影要在一个小时解决问题。这是靠场景的叠加，唤醒你头脑中的背景影像，这跟建筑是一致的。我们现在谈到的维度是时间，时间就是四维，四维并不是空间。从空间直接变换成时间这个问题，本身就是有问题的，时间还是时间。我理解四维就是空间形成，低维度的空间就是高维度的投射。我们可以在二维的纸上画出三维的图，证明二维的景象是三维投射的结果。按照这样的逻辑来讲，那三维的空间一定是四维投射的结果。我理解的四维就是人的大脑。我的建筑是存在我的头脑之中的，我将其投射到三维空间，形成了的一定是已经形成了的东西。蜜蜂在蜂巢建了之前，那个蜂巢一定是在他的思想中形成了或者是观念性的形成了，这个蜂巢我们所看到的结果，不过是蜂巢在蜜蜂头脑中投射的一个影像而已。建筑师在设计之前，这个建筑基本上是在他的头脑中形成了。

现在很多的建筑都是照片建筑，照片上好看，我希望的不是照片上的建筑，而是空间上的建筑。你要不断地运动，场景不断地展开，就有点像建筑的电影，一个一个场景地叠加。你看了一个半小时，理解了这个电影里讲的故事。你不可以仅凭一张剧照就可以理解的整部电影。实际上建筑是一个这样的形态，人在里面要游走形成一个完整的印象，那种感受才是建筑。建筑是你在头脑中重新形成的一个完整的景象，房子只不过是一个背景，所以我觉得电影和建筑就是这层的关系了。

图纸与建筑——充分或必要？

黄居正：西方中古时候的建筑就是没有图纸的。没有后来的制纸技术，书用羊皮制作，价格昂贵，活字印刷也尚未传到欧洲，一般人买不起看不起书，知识都靠修道院的僧侣或教会人士传递，所以那时没有建筑图纸也是可以想象的。如果我没记错的话，根据西方的建筑史和考古成果，雕塑和建筑没有设计图纸的存在，所有的资料都存在各个行业门类——泥瓦匠、木匠、石匠等匠人的头脑里，开工前，头人最多在沙地上的画一画，画个大概，并会告诉操作者石块怎么垒，砖块如何砌，雕刻怎么雕。直到文艺复兴早期，造纸的传入和工程师、建筑师的分工，才有图纸的出现。

王昀：这个也挺有意思的，我们现在做设计，做得好不好就看你图纸画得好不好。建筑和图纸有关系，但在某种程度上又没有太大的关系。我们做室内设计的时候会有一个问题，业主感觉你没做室内设计，就是因为文艺复兴之后的房子，室内设计是越做越复杂，都要表现在图纸上。

黄居正：建筑和图纸不是一一对应的，可能你画图纸非常好看，但建筑会非常丑陋，建筑工作除了图纸还有工地。刚才讲欧洲中古以前的建筑根本就没有图纸，各个行会的领导人在工地上告诉匠人该怎么干，建造过程既代表了工匠们的水平，最后的成果也是他们的荣誉所在，所以他们会竭尽所能，建造出许多美轮美奂的哥特式建筑。现代建筑和设计的先驱——英国工艺美术运动，其精神领袖拉斯金反对工业化，恰是因为当时的工业化产品太粗糙丑陋，所以他号召要"哥特复兴"。他还在《建筑七灯》里面提到了一个非常重要的观点，就是建筑的价值要体现在工匠的劳动上面，工匠付出了辛勤的劳动，建筑才是美的，这是建筑的伦理价值。有意思的是，他坚守的这一观点，导致了他与唯美主义画家惠斯勒之间那场艺术史上著名的官司。

文艺复兴的缺失——艺术态度的变化

王昀：实际上中世纪的很多东西还是很有魅力的，文艺复兴时期那些建筑非常繁琐。但是看中世纪的小城邦国家、古罗马建筑、古希腊的神庙，和文艺复兴时期是完全不同的。米开朗琪罗的雕塑是没有办法和古希腊的雕塑去比较的，他越做越形式化，到了罗丹，就更没有办法和古希腊的去比了，为了某种技巧去做和为了表现某种表现和状态而达到很高技巧，完全是两回事。

黄居正：拉斯金就说过，工匠是付出了他的全部感情和心血去做的。古希腊的艺术家也是在作品中倾注了一种情感，甚至是宗教式的，并非只是为了求名求利。文艺复兴时期因为阿尔伯蒂、瓦萨里、莱昂纳多·达·芬奇等人的努力，艺术家和工匠开始了分离，他们的社会地位提高到足以与人文主义学者平起平坐了，加上赞助人制度的变化、艺术市场和展览体系的建立，对待艺术的态度自然会在许多艺术家身上产生变化，某种程度上可以说是一种异化。建筑亦复如是。

柳青：但是米开朗琪罗和罗丹时代的技术也进步了。

王昀：你要是到希腊的博物馆去看雕塑海神波塞冬，会发现它特别的安静，在里面你会感觉自然有一种力量。再看米开朗琪罗，就感觉是故意了，故意表现一种力量。不经意之间表达某种力量和故意表达力量，这完全是不同层次的。这是一个需要我们这个时代思考的问题。

黄居正：这要做一种减法，是整个社会的问题。

王昀访谈 / Interview

采访人／黄元炤
北京 2011.11.9

黄元炤：在目前中国当代建筑界，聚落研究已是您个人最明显的表征，也成为您个人对外表述的形象语言。能不能谈谈当时聚落研究如何勾起您感兴趣的点？之后为何钟情于聚落研究？

王昀：实际上在20世纪的80年代，国内曾经掀起一股民居研究热潮，在我毕业设计时，是周人忠老师带我们去参观实习的。按照当时的习惯应该去广州、深圳去看那个时代的新建筑，但周老师说那些地方什么时候都可以去的，建议我们去乡下去看看。于是我们就去了云南丽江，那是1985年初，当时的丽江没有对外开放，外国人只能到大理不能去丽江，记得我们公共汽车坐了两天，夜里在中途的一个地方还睡了一夜，具体是什么地方我忘了。但是到丽江看了以后，感觉丽江古镇特别有味道，感觉一个一个的建筑连在一块儿的状态特别棒，但是那个时候对于聚落的概念应该还没那么清晰，回来后我写了一篇文章，"四合院建筑型制的同构关系初探"，实际上就是看了丽江的一个院落，发现了那个里面主人所写的自宅中的对联，对四个不同朝向房屋的意境描述，象征着春夏秋冬，表示了一种宇宙意境，回来后查了一些资料，写了这篇文章。

到日本以后，当时东京大学的原藤井研究室是研究聚落最有权威性的研究室，我个人也很想知道这个研究室对民居的研究有什么样的看法？有怎样的分析？最后就到了这个研究室，但我去的时候，他们之前已经有很长时间不做聚落研究了，他们那时在做城市研究。因为原广司老师那时已经忙起来了，他已成为当时日本最受瞩目的建筑大师，他的工程项目不断，离不开东京，于是他就开始关注城市了。1993年、1994年以后，倒是藤井明老师又重新做聚落调查，因为藤井明老师最早在20世纪70年代就参与到第一个聚落调查，几乎每一次聚落调查他都去过。

我到研究室的时候，研究室正在关心一件事，就是如何对聚落进行全面的梳理与解析，我的研究工作就是把聚落整体的平面图进行数字化，输入到计算机里面，然后建立数理模型，编计算机程序进行分类，而这就成为我最后的博士论文。

黄元炤：您刚才说到，聚落研究跟民居在研究的视点上还是有很大的不同，其实在中国近代的建筑发展当中，关于探访与参访的部分，一开始都是锁定在古建筑与民居，这个跟您所关注的聚落是不一样的。我做了一个梳理，比如说刘敦桢，1934年与夏昌世、梁思成、卢树森去测绘苏州古建筑。杨廷宝，1973年赴山西五台及雁北地区勘查古建筑。辜其一，重庆建工学院古建专家，考察过宋代建筑，三国姜维民俗和芦山庆坛、花灯等。徐尚志，对少数民族建筑和各地民居进行长期的调查研究。王翠兰，考察与探访西南贫困边远的少数民族民居，主持云南民居调查等。可见，在中国近代的建筑发展中，关于探访与参访的方面，与您所关注的聚落实然不同，我想听听您进行聚落探访的过程与目的，与这些人有哪些差异？

王昀：我最近写了一本《向世界聚落学习》的书，在写的过程当中，确实思考了很多东西，就是你刚才提的问题，我曾经考虑过，就是为什么我们过去叫民居研究，民居研究和聚落研究有什么不一样呢？对象物上都是一样的，民居也是在村落当中，然后聚落，实际上是民居一个整体的、集合的呈现，它的对象还是民居，那为什么我不叫民居研究，一定要叫聚落研究，要知道聚落这个词并不是这几年发明的，很早就有，聚在一块的民居就是聚落。

黄元炤：那有什么不一样呢？

王昀：我认为还是一个视点的不同，我一直想强调的就视点，就是大脑当中所要期待看到的那个东西是什么，很重要，并不是我长了眼睛，眼睛就可以发现东西，而是说我大脑后面想要看到的是什么东西，它才可以看到。最简单的例子就是画水彩画，画阴影的时候，绿色的叶子加一点红，那个红色你是看不到的，你看的是一个绿叶，那个红色是经过分析后得到的，但是你若说阴影里面有红色的时候，你再去看阴影就可以看到红色了。我们过去看民居的时候，如果还是希望在民居当中找到一个优秀的房子，地主家或者说有钱的人家的豪宅，其实这种做法跟传统的建筑史教育是一样的，虽然进到了村落中，但是脑子里还是想找到一个村落中的宫殿。

如果用远景、中景和近景来观察，民居是一个近景的研究，而聚落是一个远景的研究，放到远景以后呢，很多事情都模糊了，模糊之后，某些细节不见得看得那么细，但是会瞬间在整体上能够进行一个把握，我认为这一点是聚落研究跟民居研究不一样的地方。

黄元炤：所以，您的设计创作也一直处于近景和远景之间在进行着。就我的观察，您其实一直游走于传统与现代之间，您所关注的传统是原生聚落，而您所关注的现代是现代主义建筑，这两者在建筑、历史、文化、类型、语言方面都有极大的差异性及非等同性，您也似乎想要模糊这两者，企图让自己处于一个错乱或迷乱的状态，且将传统与现代这两者共同呈现和存在于您的建筑当中，您的作品可以说是现代主义，又可以说是聚落体验后的再现与重现，这部分您的看法如何？

王昀：你说错乱的状态？

黄元炤：也就是一种模糊的状态。

王昀：确切一点说还是错乱。这么一说，我从小的经历还真是有些错乱和漂浮，总换地方。搬到这儿，然后又搬到那儿，从哈尔滨搬到北京来，到处看。学了建筑之后也到处乱看，看完了以后，又上日本待了很长的时间，一会儿看这个，看那个，又看建筑，又看聚落，所以，想我这个人肯定就是错乱和混乱了。错乱完了以后，我在做设计时，毕竟没办法错乱了，当我拿出一个东西给你看的时候，一定是我把所有错乱的集合体压缩了以后，拿出来的一件东西，可能你看的东西很简单，但是这个简单的背后，是因为有过很多的积累后的一个结果，所以我更愿意说，建筑是设计师的一个概念的投射结果。

为什么讲投射，就是说我脑子里面的那个东西和这个东西是对应的，但是它们两个没有必然的牵扯和联系，比方说我做茶杯，在我脑子里面，这个茶杯的形象和现实间的关系是一个对应的、投射的关系，它在我脑子里的那个点，放在现实当中，变成一个具象物，设计是这样的，建筑设计也是如此。

我做的东西，事实上是我所有经历总和的一个结果，浓缩成一个东西拿出来的。我一直在想，我怎么能够让建筑做得没有明显的特征，就是说让建筑直接看不出我的经历，因为我在北京、哈尔滨、东京都生活过，也到过非洲和欧洲，哪儿都转悠过了，都影响了。设计时做什么风格呢？做不出杂乱的东西，就只好是交白卷了，交一张白纸，做一个白盒子，可能是最简单的，可能是最不错乱的东西。

黄元炤：其实我观察到的也是一种摇摆的状态，一下子这样，一下子那样。有时候可能是这个比较多一点，那个比较多一点，有一种摇摆。其实很多人看您的作品很直观的觉得跟现代主义有点关系，那么您自己对外传达说，这个是经过我聚落体验后的重现和再现，这偏向于一种转化的说法，但外人看还是一个现代主义的建筑，就是您这个主体与其他人的这个客体，共同对您所做出来的这个对象物，关注与切入的视点是不同的，也许有的人理解您所说的聚落，会更贴近于您的想法，但有的人不理解的话，看来看去还是一个现代主义的东西。所以若客观来看，就是您会给人有一种摇摆的感觉，这部分您有什么看法？

王昀：如何可以把自己思考的极限推到一个什么样的边缘，这是一个重要的事。而这一点，我认为现代主义的起始点并不是因为一个造型，而是它把人的理性思考做到了一个纯粹的极致化，20世纪初大家做的，也就是艺术最边缘的一件事，就是抽象可以抽象到什么程度，我认为这件事是需要我们考虑的，实际上今天面临的问题也是一样的，就是说我们的建筑可以抽象到一个什么样的程度，这件事如果我们不去探索的话，事实上你下面往回捣的那个工作，不见得可以那么顺利，不见得可以捣腾过来，有的时候在混沌当中就被搅和进去了，我想有更多建筑师捣腾这件事的时候，就是为了把这件事捣腾清楚。

黄元炤：就我的观察，北京庐师山庄会所与别墅，您是在现代主义基础上关注的空间，有种空间抽象的体验性，利用抽象表述出作品内在深层的真实性，在几何体块中，创造出一种内部游走与想象的空间体验过程，形塑出多重视点、多重形式的转换规则，与空间的多重发展，有一种由内向内看的随机指向性。然而在作品中，当空间抽象与体验发挥到极致的时候，您的建筑师角色在建筑空间中顿时地抽离、消失、隐藏或退位，并与观察者同时成为第三人称，也成为一个赏目者，而空间本体则跃升为第一人称，成为主角，并展示内部富有情结般的路径，您对这部分有何看法？

王昀：实际上是这样的，建筑是设计师的一个概念的投射，投射出来结果以后，它就变成了对象物，你刚才讲了建筑空间性，可是那个空间，人家是看不见的，空间是需要有限定形成的，那么你怎么样去限定它，而这个空间的组合关系其实来得更重要，用什么样的顺位把它组合起来，什么地方扩大，什么地方缩小，是因设计师的不同而不同的，像庐师山庄建成后，确实不同的人进去了以后，都会唤起他们的某一些记忆，这件事是我感兴趣的。

尽管这个设计是依照我的经验，并且那个房子是很抽象的，但是不同人走到这样一个空间当中，可以把儿时的记忆，或者说小时候经历过的一些事在里面调动起来的话，我认为这个东西是有意义的，而这才是建筑最本质的一个东西。

黄元炤：您的空间其实想要去刺激人达到一种想象，我觉得这个想象跟动作有一点关系。体验的过程，也许一开始是缓慢的，然后可以从缓慢到停留，或者说在停留的时候，可以关注到这个场景与下一个场景之间的衔接，或者说去品位空间的氛围，然后当人们停留了以后，脑袋中的想象强度会不断地扩大和向往，就是会忆到设计师想要表达的意识空间，就是达到来回反复的意识讯息的接触。所以，您的作品当中，常常会有一些坡道的应用，或者说将行进路径拉长，我觉得就是人行进过程中要制造缓慢一点，可是这个缓慢的过程，在动的时候，他的想象也许是不全的，也许是一半的，可是当过道拉长了以后，到达了一个平台，人处于停留状态后，这个想象也许是全的。所以，您的设计在塑造一种路径，而这个路径是从一种缓慢到停留再到缓慢，来来回回重复着，是从缓慢而形成的半想象，到停留而形成的全想象，这部分您有何看法？

王昀：我个人理解就是建筑设计怎么样做得好玩，你刚才说的这个东西来得很重要，就是空间怎么设定它。其实设计师，就是像北京人常说的：你又不是别人肚里的蛔虫，你怎么知道别人怎么想的。而相反的，你是根据你自己的经验来判断设计的，这是因为你也是人，中国有句名言："老吾老以及人之老，幼吾幼以及人之幼。"人们都是从自我的角度去思考，去判断别人对这个东西的理解，作为设计师重要的是挖掘你自身，因为你也是人，你认为这个东西好，你才去进行设计，对吧！

实际上在做设计的时候，情景判断，或者说空间的整体思考很重要。因为建筑不同于绘画，也不同于雕塑。作为雕塑，我从一个角度看就可以了，或者说现在雕塑也讲时间的概念，融合进去。建筑更是，建筑不是一个造型的融合，更重要的是房间里面，门厅进去后的感受，怎么通过一个狭窄的地方，怎么走，怎么样到房间，家具布置是什么样子，所有的一切都和空间发生关系，空间这个词是看不见、摸不到的，是需要墙面这些东西来做的。

那么对空间路线的设定呢，就是像你说的，什么地方用坡道，让它延缓；什么地方用楼梯，迅速把这个问题解决。实际上这些事呢，是跟要表达的整体印象是有关的，其实我一直认为建筑最后不是看一张照片和一个场景，而是你走进去了，转完了以后所获得的整体场景的总和，才是建筑最终要感受到的东西，这个是和电影、戏剧有关的。电影不可能靠一个剧照就把电影表达清楚，必须坐在那儿看120分钟，85分钟，84分钟或者165分钟，而人在建筑空间里游走的路径，就像镜头需要设定一样，一点一点地展看，就像是看了一部电影或者说看了一部戏剧一样，最后得到一个总体的感受。但是怎么样让人生活在里面有意思？有一句话，叫历久弥新，就是时间越久，就感觉很多新的东西在里面，实际上这个里面有很多的设定是需要来做的。

黄元炤：我观察您的作品还是有一个一致性，比如说从庐师山庄会所与住宅，到百子湾中小学，再到近期的杭州西溪湿地创意会所……就是说客观地观察，假如说我们不进入这个空间，从外部看整体，还是带有一种一致性，要进入空间，才能更深入去体验。可是这个一致性，若回到建筑学的观点，先不谈聚落体验的一些重现，它其实是带有纯粹、纯净的设计倾向，就是现代主义所讲的纯粹性。但是您好像又尝试了不同的方向，如北京石景山财政中心。就我的观察，这个项目在一个几何形体的基础上，您想体现一种建筑的巨大化，一种象征性的精神表述，比如说入口有巨大尺度的虚框，后面是巨大的几何实体，似乎想做出一个城市空间中的标志，这个部分您自己有什么看法呢？

王昀：设计的时候，环境是挺重要的，比如说像你刚才前面说的，像庐师山庄，包括后来的都是小房子，那么到城市化的一个大房子的时候，如何可以让它感觉到是符合那个尺度和环境呢？就说石景山财政局这个建筑吧。因为那个环境用地非常的狭窄，左边是法院，已经盖好了，右边是巨大的商场。一般的政府的办公楼，给人的形象总是一个威严的状态，我想能不能财政局盖完了以后，让人感觉是比较亲切一点，有一个接纳的状态，用框呈现出接纳的状态。而这个呈现接纳状态的框，在城市中又是大尺度的，我想让这个框下面的台阶在夏天坐满周边的居民，让这个场所成为一个面向城市的广场与舞台，上面的大框更有一个大舞台的台口意味。

黄元炤：最主要就是您做了那个框，那个框的尺度在现场，给人巨大的感觉。

王昀：那个框还是比较薄、比较细的，每一个人的解释是不一样的，我认为这还是立场的问题，当你把政府看成是一个衙门的时候，它就威严起来，你把它看成是一个大的剧场，大家都可以在这儿集合，到这来聚集，它就是一个广场，就是一个公共的小剧场，这个事儿还是立场的问题，就是你去看有一个俄罗斯的小说《安娜·卡列尼娜》，对这个小说的解读是多面的，从一个革命者的角度来讲，主人公是冲破旧婚姻的一个女性代表，那从一个保守主义来看，那就是一个荡妇。

所以我一直认为建筑去解读它的时候，立场问题太重要了，这让我回想到20世纪80年代，当时我们去解读现代主义建筑的时候，说那是没有人性的东西，住宅是居住的机器，当时国内很有名的学者都到处去批判，也说后现代是多有人性，现在当立场一变了以后，后现代又说成是浪费。过几天，可能我们今天的很多建筑，你认为又是泡沫经济的产物，又是一个浪费的消费产物，那么怎么解释这件事呢？在当今这个时代看，一些建筑似乎象征了我们所处的现状，明天一转身泡沫经济破灭，这一切可能变成了泡沫建筑，变成泡沫时代的建筑设计，泡沫风格，所以我认为设定立场和看问题的角度，让对建筑的判断更加客观一点，是摆在我们面前一个很重要的责任和话题。

黄元炤：杭州西溪湿地创意会所，算是您近期最新的一件作品，在这个创意会所当中，您提到了一种"共同幻想"的表述，一种离散式聚落的概念，我的观察，其实这个会所就是您多年来一个设计的总结。

王昀：可能是转型的开始？

出版物一览
Publication List

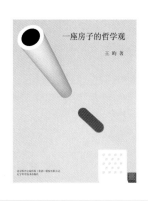

书　　名：从风景到风景	书　　名：一座房子的哲学观	书　　名：传统聚落结构中的空间概念	书　　名：空间的界限
作　　者：王昀	作　　者：王昀	作　　者：王昀	作　　者：王昀
开　　本：16开	开　　本：24开	开　　本：16开	开　　本：32开
页　　数：166	页　　数：258	页　　数：484	页　　数：181
装　　帧：平装	装　　帧：平装	装　　帧：平装	装　　帧：平装
出 版 社：中国电力出版社	出 版 社：辽宁科学技术出版社	出 版 社：中国建筑工业出版社	出 版 社：辽宁科学技术出版社
出版年月：2010.1	出版年月：2010.1	出版年月：2009.8	出版年月：2009.9

书　　名：空间穿越
作　　者：王昀
开　　本：32开
页　　数：266
装　　帧：平装
出 版 社：辽宁科学技术出版社
出版年月：2010.8

书　　名：空谈空间
作　　者：王昀
开　　本：16开
页　　数：223
装　　帧：平装
出 版 社：辽宁科学技术出版社
出版年月：2010.8

书　　名：空间的潜像
作　　者：王昀
开　　本：32开
页　　数：323
装　　帧：平装
出 版 社：中国建筑工业出版社
出版年月：2011.4

书　　名：向世界聚落学习
作　　者：王昀
开　　本：32开
页　　数：219
装　　帧：平装
出 版 社：中国建筑工业出版社
出版年月：2011.9

书　　名：跨界设计——建筑与音乐
作　　者：王昀
开　　本：16开
页　　数：108
装　　帧：平装
出 版 社：中国电力出版社
出版年月：2012.1

书　　名：向世界聚落学习
作　　者：王昀
开　　本：12开
页　　数：269
装　　帧：精装
出 版 社：中国建筑工业出版社
出版年月：2012.1

作品年表
Chronology of Works

范例
1 建筑名称
2 设计者
3 结构设计
4 所在地
5 设计时间
6 施工时间
7 占地面积
8 建筑面积

▲——收录作品

1 建筑名称：60平方米极小城市 ▲
2 设计者：王昀
3 结构设计：——
4 所在地：北京海淀区北京大学燕北园
5 设计时间：2002年5月
6 施工时间：2002年11月
7 占地面积：60平方米
8 建筑面积：60平方米

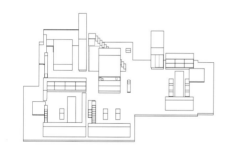

1 建筑名称：善美办公楼门厅增建 ▲
2 设计者：王昀
3 结构设计：马亚林
4 所在地：北京市朝阳区东方东路11号
5 设计时间：2002年6月
6 施工时间：2003年4月
7 占地面积：170平方米
8 建筑面积：92平方米

1 建筑名称：建邦会馆
2 设计者：王昀
3 结构设计：——
4 所在地：北京市西城区展览路
5 设计时间：2002年6月
6 施工时间：2003年4月
7 占地面积：2420平方米
8 建筑面积：22000平方米

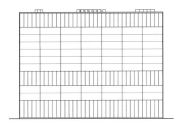
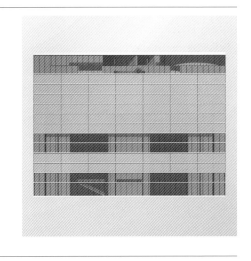

1 建筑名称：Q宅
2 设计者：王昀
3 结构设计：李东升
4 所在地：北京市怀柔区东凤山
5 设计时间：2003年3月
6 施工时间：2003年10月
7 占地面积：600平方米
8 建筑面积：410平方米

1 建筑名称：庐师山庄A+B住宅 ▲
2 设计者：王昀
3 结构设计：黄宇
4 所在地：北京市石景山区八大处庐师山庄
5 设计时间：2003年3月
6 施工时间：2005年10月
7 占地面积：1380平方米
8 建筑面积：A：650平方米 B：640平方米

1 建筑名称：庐师山庄会所 ▲
2 设计者：王昀
3 结构设计：黄宇
4 所在地：北京市石景山区八大处庐师山庄
5 设计时间：2003年5月
6 施工时间：2005年10月
7 占地面积：790平方米
8 建筑面积：1600平方米

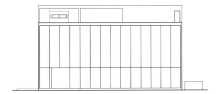

1 建筑名称：百子湾小区幼儿园 ▲
2 设计者：王昀
3 结构设计：李东升
4 所在地：北京市朝阳区百子湾路1号
5 设计时间：2003年6月
6 施工时间：2006年11月
7 占地面积：1570平方米
8 建筑面积：3200平方米

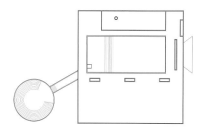

1 建筑名称：百子湾小区中学 ▲
2 设计者：王昀
3 结构设计：季克平
4 所在地：北京市朝阳区百子湾路1号
5 设计时间：2003年6月
6 施工时间：2006年11月
7 占地面积：4380平方米
8 建筑面积：11000平方米

1 建筑名称：百子湾小区车库
2 设计者：王昀
3 结构设计：季克平
4 所在地：北京市朝阳区百子湾路1号
5 设计时间：2003年5月
6 施工时间：——
7 占地面积：2820平方米
8 建筑面积：12700平方米

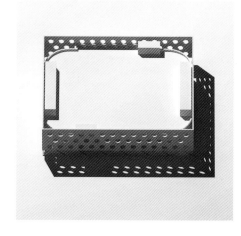

1 建筑名称：石景山财政培训中心 ▲
2 设计者：王昀
3 结构设计：郝卫清、张富昌
4 所在地：北京市石景山区阜石路167号
5 设计时间：2003年9月
6 施工时间：2007年8月
7 占地面积：2400平方米
8 建筑面积：7000平方米

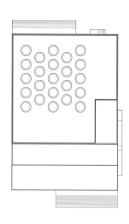

1 建筑名称：河南科技大学校园规划
2 设计者：王昀
3 结构设计：——
4 所在地：河南省洛阳市
5 设计时间：2003年5月
6 施工时间：——
7 占地面积：2442660平方米
8 建筑面积：1882260平方米

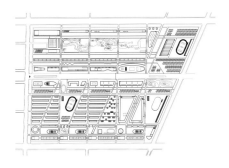
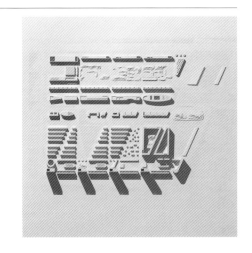

1 建筑名称：善美售楼处
2 设计者：王昀
3 结构设计：——
4 所在地：北京市朝阳区
5 设计时间：2003年8月
6 施工时间：——
7 占地面积：880平方米
8 建筑面积：490平方米

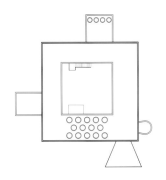
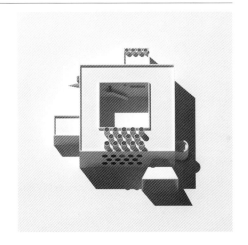

1 建筑名称：柳州电器厂住宅
2 设计者：王昀
3 结构设计：季克平
4 所在地：广西壮族自治区柳州市
5 设计时间：2004年3月
6 施工时间：——
7 占地面积：280平方米
8 建筑面积：1400平方米

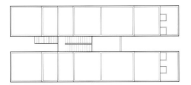

1 建筑名称：共同体住宅
2 设计者：王昀
3 结构设计：——
4 所在地：北京市昌平区
5 设计时间：2004年5月
6 施工时间：——
7 占地面积：2045平方米
8 建筑面积：1307平方米

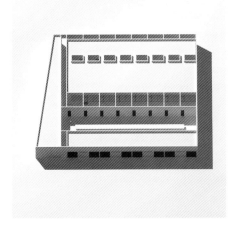

1 建筑名称：石景山财政局车库入口
2 设计者：王昀
3 结构设计：郝卫清、张富昌
4 所在地：北京市石景山区阜石路167号
5 设计时间：2005年3月
6 施工时间：——
7 占地面积：310平方米
8 建筑面积：560平方米

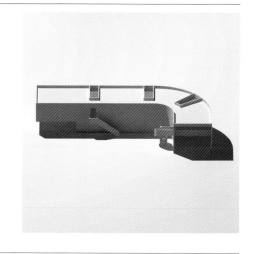

1 建筑名称：大舜天成尚都综合楼
2 设计者：王昀
3 结构设计：李东升
4 所在地：山东省济南市解放路
5 设计时间：2005年4月
6 施工时间：——
7 占地面积：4837平方米
8 建筑面积：17500平方米

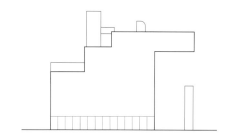
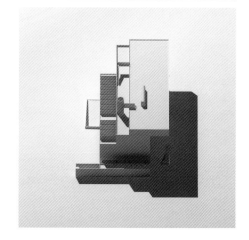

1 建筑名称：国轩国际研发中心商业设施 ▲
2 设计者：王昀
3 结构设计：宗国华
4 所在地：北京市朝阳区北埠东郊农场路
5 设计时间：2005年3月
6 施工时间：——
7 占地面积：4890平方米
8 建筑面积：16120平方米

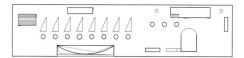

1 建筑名称：吴家场经济适用房住宅小区
2 设计者：王昀
3 结构设计：井荣恩、李影
4 所在地：北京市海淀区
5 设计时间：2005年3月
6 施工时间：2011年
7 占地面积：37000平方米
8 建筑面积：126125平方米

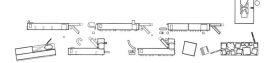

1 建筑名称：吴家场小区社区服务中心
2 设计者：王昀
3 结构设计：——
4 所在地：北京市海淀区
5 设计时间：2005年3月
6 施工时间：——
7 占地面积：932平方米
8 建筑面积：2900平方米

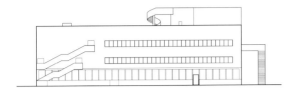
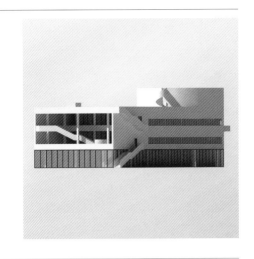

1 建筑名称：吴家场小区幼儿园
2 设计者：王昀
3 结构设计：——
4 所在地：北京市海淀区
5 设计时间：2005年3月
6 施工时间：——
7 占地面积：1139平方米
8 建筑面积：2260平方米

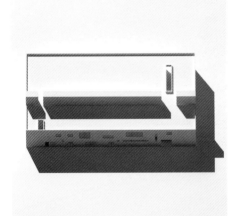

1 建筑名称：欧陆经典立面改造
2 设计者：王昀
3 结构设计：——
4 所在地：北京市朝阳区
5 设计时间：2005年6月
6 施工时间：2006年11月
7 占地面积：8239平方米
8 建筑面积：23376平方米

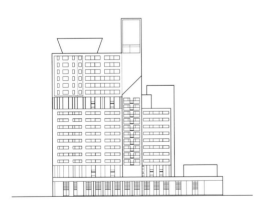

1 建筑名称：long宅
2 设计者：王昀
3 结构设计：田俊杰
4 所在地：北京市亦庄
5 设计时间：2007年11月
6 施工时间：
7 占地面积：27665平方米
8 建筑面积：1600平方米

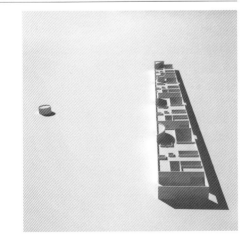

1 建筑名称：太阳村
2 设计者：王昀
3 结构设计：季克平
4 所在地：陕西省陇县
5 设计时间：2006年
6 施工时间：——
7 占地面积：10500平方米
8 建筑面积：1755平方米

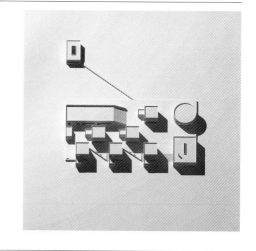

1 建筑名称：茵莱玻璃钢门窗制品有限公司办公楼 ▲
2 设计者：王昀
3 结构设计：季克平
4 所在地：北京市通州区
5 设计时间：2006年7月
6 施工时间：2007年12月
7 占地面积：590平方米
8 建筑面积：1580平方米

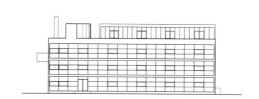

1 建筑名称：济南大舜天成青年公寓
2 设计者：王昀
3 结构设计：李东升
4 所在地：山东省济南市
5 设计时间：2009年3月
6 施工时间：——
7 占地面积：11200平方米
8 建筑面积：50108平方米

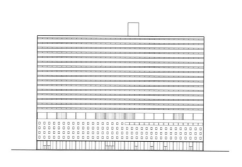
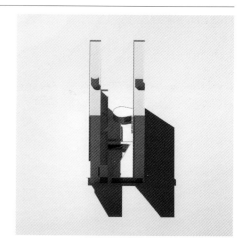

1 建筑名称：内蒙古库伦中学
2 设计者：王昀
3 结构设计：文辉
4 所在地：内蒙古自治区呼和浩特市
5 设计时间：2007年5月
6 施工时间：2009年9月
7 占地面积：29320平方米
8 建筑面积：12440平方米

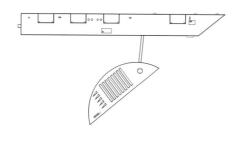
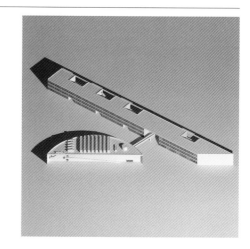

1 建筑名称：无锡天鹅湖会所A ▲
2 设计者：王昀
3 结构设计：李东升
4 所在地：江苏省无锡市
5 设计时间：2007年
6 施工时间：——
7 占地面积：819平方米
8 建筑面积：1000平方米

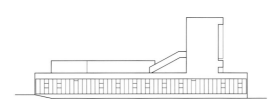
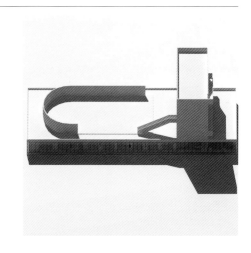

1 建筑名称：无锡天鹅湖会所B ▲
2 设计者：王昀
3 结构设计：李东升
4 所在地：江苏省无锡市
5 设计时间：2007年
6 施工时间：——
7 占地面积：835平方米
8 建筑面积：1000平方米

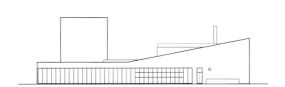
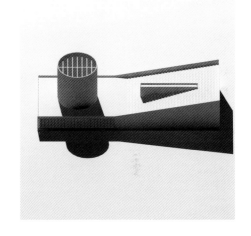

1 建筑名称：妫水创意酒店
2 设计者：王昀
3 结构设计：李克平
4 所在地：北京市延庆县
5 设计时间：2007年2月
6 施工时间：——
7 占地面积：16310平方米
8 建筑面积：27940平方米

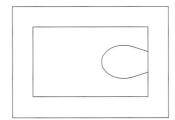
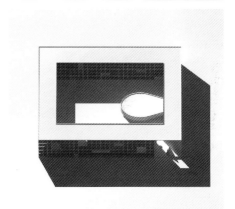

1 建筑名称：北京万象新天150平方米住宅改造 ▲
2 设计者：王昀
3 结构设计：——
4 所在地：北京市朝阳区
5 设计时间：2007年1月
6 施工时间：2007年7月
7 占地面积：
8 建筑面积：150平方米

1 建筑名称：西溪学社 ▲
2 设计者：王昀
3 结构设计：李东升
4 所在地：浙江省杭州市余杭区
5 设计时间：2007年10月
6 施工时间：2010年1月
7 占地面积：59830平方米
8 建筑面积：3800平方米

1 建筑名称：深圳南油广场
2 设计者：王昀
3 结构设计：——
4 所在地：深圳市
5 设计时间：2008年4月
6 施工时间：——
7 占地面积：171070平方米
8 建筑面积：456670平方米

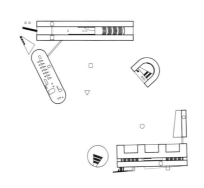
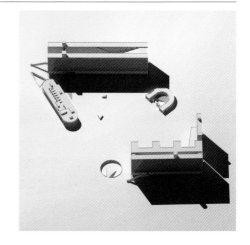

1 建筑名称：柔软住宅 ▲
2 设计者：王昀
3 结构设计：李东升
4 所在地：北京市亦庄
5 设计时间：2008年4月
6 施工时间：
7 占地面积：1400平方米
8 建筑面积：780平方米

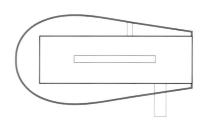
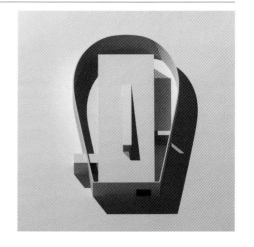

1 建筑名称：诸暨商业文化街
2 设计者：王昀
3 结构设计：李东升
4 所在地：浙江省诸暨市
5 设计时间：2008年7月
6 施工时间：——
7 占地面积：687750平方米
8 建筑面积：716895平方米

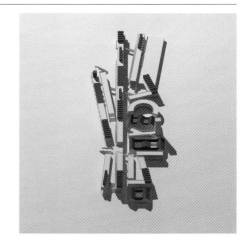

1 建筑名称：苏家坨住宅区小学 ▲
2 设计者：王昀
3 结构设计：李东升
4 所在地：北京市海淀区
5 设计时间：2009年8月
6 施工时间：——
7 占地面积：10100平方米
8 建筑面积：4969平方米

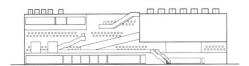
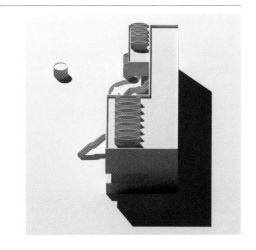

1 建筑名称：欧陆经典会所改造
2 设计者：王昀
3 结构设计：李东升
4 所在地：北京市朝阳区
5 设计时间：2009年2月
6 施工时间：——
7 占地面积：1818平方米
8 建筑面积：5161平方米

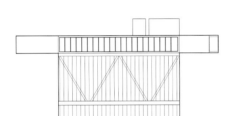

1 建筑名称：鄂尔多斯康巴什第六中学校设计 ▲
2 设计者：王昀
3 结构设计：——
4 所在地：内蒙古康巴什
5 设计时间：2010年9月
6 施工时间：——
7 占地面积：6500平方米
8 建筑面积：20743平方米

1 建筑名称：私宅 ▲
2 设计者：王昀
3 结构设计：李影
4 所在地：浙江省杭州市
5 设计时间：2010年6月
6 施工时间：——
7 占地面积：2030平方米
8 建筑面积：454平方米

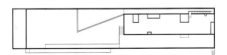
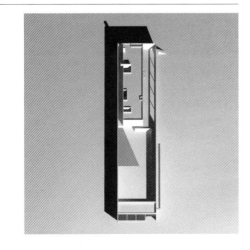

1 建筑名称：ERDOS P23
2 设计者：王昀
3 结构设计：季克平
4 所在地：内蒙古东胜区
5 设计时间：2010年1月
6 施工时间：2011年7月
7 占地面积：4456平方米
8 建筑面积：8370平方米

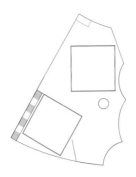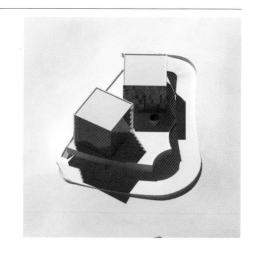

1 建筑名称：ERDOS T15
2 设计者：王昀
3 结构设计：季克平
4 所在地：内蒙古东胜区
5 设计时间：2010年1月
6 施工时间：2011年7月
7 占地面积：2122平方米
8 建筑面积：10998平方米

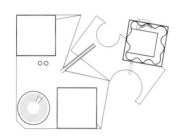

1 建筑名称：鄂尔多斯康巴什住宅组团设计 ▲
2 设计者：王昀
3 结构设计：——
4 所在地：内蒙古康巴什
5 设计时间：2011年2月
6 施工时间：——
7 占地面积：14520平方米
8 建筑面积：7400平方米

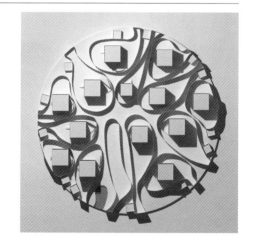

1 建筑名称：铜川城市展览馆
2 设计者：王昀
3 结构设计：——
4 所在地：陕西省铜川市
5 设计时间：2011年10月
6 施工时间：——
7 占地面积：4896平方米
8 建筑面积：13550平方米

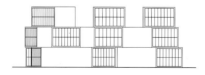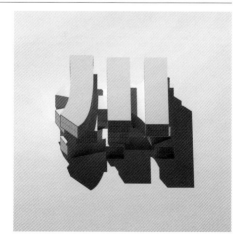

1 建筑名称：吴家场小区幼儿园
2 设计者：王昀
3 结构设计：季克平
4 所在地：北京市海淀区
5 设计时间：2011年11月
6 施工时间：——
7 占地面积：994平方米
8 建筑面积：2260平方米

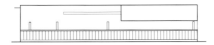
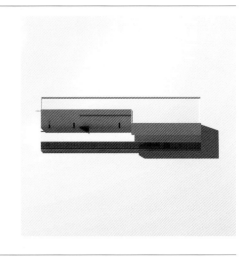

1 建筑名称：吴家场小区文体活动中心
2 设计者：王昀
3 结构设计：季克平
4 所在地：北京市海淀区
5 设计时间：2011年11月
6 施工时间：——
7 占地面积：1085平方米
8 建筑面积：1760平方米

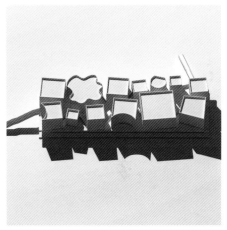

本书参编人员：赵冠男、张捍平、程艳春、张振坤

曾参与设计的工作室成员（按姓氏笔画）

丁　晨、马　磊、王　涛、王东蒙、孔梦冉、过海光、
刘欣鹏、庄浩然、苏　斯、苏立恒、李　图、李　楠、
李　婉、李　喆、杨　晨、杨笑北、吴燕文、张　扬、
张　然、张文贺、张振坤、张捍平、张博强、邵　鹚、
周　毅、周静敏、赵　凡、赵　佳、赵冠男、赵超超、
赵晓博、禹　航、俞文婧、高　冬、郭　岩、郭　婧、
黄　玉、黄　吉、黄　威、黄晓殊、曹明梁、章跃麒、
寇佳意、程艳春、樊友广

项目施工图合作单位：
北京建筑工程设计公司；
北京互联盟建筑设计有限公司；
北京马建国际建筑设计顾问有限公司；
北京筑都方圆建筑设计有限公司

协作建筑师：柯伟来、吉树起、王汝峰

王昀简介

王昀 博士

1985 年　毕业于北京建筑工程学院建筑系获学士学位

1995 年　毕业于日本东京大学获得工学硕士学位

1999 年　于日本东京大学获得工学博士学位

2001 年　执教于北京大学

2002 年　成立方体空间工作室

建筑设计竞赛获奖经历：

1993 年日本《新建筑》第 20 回日新工业建筑设计竞赛获二等奖

1994 年日本《新建筑》第 4 回 S×L 建筑设计竞赛获一等奖

主要建筑作品：

善美办公楼门厅增建，60 平方米极小城市，石景山财政局培训中心，庐师山庄，百子湾中学校，百子湾幼儿园，杭州西溪湿地艺术村 H 地块会所等

参加展览：

2004 年 6 月参加"'状态'中国青年建筑师 8 人展"

2004 年首届中国国际建筑艺术双年展参展

2006 年第二届中国国际建筑艺术双年展参展

2009 年参加在比利时布鲁塞尔举办的"'心造'——中国当代建筑前沿展"

2010 年参加威尼斯建筑艺术双年展，德国 karlsruhe Chinese Regional Architectural Creation 建筑展

2011 年参加捷克 prague 中国当代建筑展，意大利罗马"向东方－中国建筑景观"展，中国深圳·香港城市建筑双城双年展等

Profile

Dr. Wang Yun

Graduated with a Bachelor's degree from the Department of Architecture at the Beijing Institute of Architectural Engineering in 1985.

Received his Master's degree in Engineering Science from Tokyo University in 1995.

Received a Ph.D. from Tokyo University in 1999.

Taught at Peking University since 2001.

Founded the Atelier Fronti (www.fronti.cn) in 2002.

Prize:

Received the second place prize in the "New Architecture" category at Japan's 20th annual International Architectural Design Competition in 1993

Awarded the first prize in the "New Architecture" category at Japan's 4th S×L International Architectural Design Competition in 1994

Prominent works:

ShanMei Office Building Foyer, a small city of 60 square meters, the Shijingshan Bureau of Finance Training Center, Lushi Mountain Villa, Baiziwan Middle School, Baiziwan Kindergarten, and block H of the Hangzhou Xixi Wetland Art Village.

Exhibitions:

The 2004 Chinese National Young Architects 8 Man Exhibition, the First China International Architecture Biennale, the Second China International Architecture Biennale in 2006, the "Heart-Made: Cutting-Edge of Chinese Contemporary Architecture" exhibit in Brussels in 2009, the 2010 Architectural Venice Biennale, the Karlsruhe Chinese Regional Architectural Creation exhibition in Germany, the Chinese Contemporary Architecture Exhibition in Prague in 2011, the "Towards the East: Chinese Landscape Architecture" exhibition in Rome, and the Hong Kong-Shenzhen Twin Cities Urban Planning Biennale

www.fronti.cn